THE SHOTGUN CONSERVATIONIST

To my dad and the Oldfields School, always in my corner.

Photo and illustration credits appear on page 000.

Published in 2023 by Timber Press, Inc.

Hachette Book Group

1290 Avenue of the Americas

New York, NY 10104

timberpress.com

Timber Press, Inc., is a subsidiary of Workman Publishing Co., Inc.,
a subsidiary of Hachette Book Group, Inc.

Printed in the US on responsibly sourced paper.

Jacket and text design by Vincent James

ISBN 978-1-64326-103-4

A catalog record for this book is available from the Library of Congress.

THE SHOTGUN CONSERVATIONIST

Why Environmentalists Should Love Hunting

BRANT MACDUFF

TIMBER PRESS
Portland, Oregon

While my interest in natural history has added very little to my sum of achievement, it has added immeasurably to my sum of enjoyment of life.

–Theodore Roosevelt, "My Life as a Naturalist," *American Museum Journal*, 1918

Contents

Introduction

I'm not trying to turn anyone into a hunter, but I am hoping I can broaden some perspectives as to what being environmentally and animal friendly looks like. I hope the science, history, economics, and passion for animals that drove my personal journey from anti-hunting to hunter can provide you with a new perspective in less time than it took me to earn mine. Glad to have you.

To those both curious and ambivalent about hunting: If you compost, drive a hybrid, enjoy being outdoors, get your groceries from a farmers market, pride yourself on buying regeneratively raised, 100-percent-grass-fed beef or bison and sustainably sourced wild-caught Alaskan seafood, want to be more self-sufficient, and love animals, then you're in the right place. You are primed to become a hunter or hunting advocate. You give a shit about the environment, and you want your personal actions to reflect that. You're ready for the pinnacle in environmentally friendly protein sourcing—procuring your own sustainable wild meat. I think wild meat tastes better, but that's subjective. What's not subjective is how the environments it comes from are better too. Welcome.

To those who hate hunting: It's okay if you do, but I don't think you would if you knew more about it. I hated hunting for much of my life, and it wasn't a potion or spell or cult leader who changed my mind—it was me. I changed my mind about something that I thought was a core belief. You probably hate certain people who hunt because they're either gross to you in one way or another or they're terrible representatives of the lifestyle (and most likely the only members of the community you hear about—because who ever hears about nice people doing things by the book?). But why do you hate *hunting*? Because someone killed an animal and you like animals? That's basically how I felt, but I learned that the math of caring about animals and the environment isn't so easy. And I still love animals.

To those who hate hunting and hunters and also eat meat: I was like that once. But the sheer crushing weight of the hypocrisy finally became too much for me. I loved animals, but I was eating them; I had heard commercially produced meats were terrible for the environment, but I still bought them–somehow, I saw the duck hunter as flawed, while my basket of twenty-five-cent wings ranneth over. It's important to remember that just because you didn't shoot the rotisserie chicken at the grocery store, you're not any less culpable for its death. In hunting, I discovered (in this thing that's been around for forever) a way to eat meat and give back to the earth while becoming more connected to it at the same time. It's not that I wanted to shoot chickens myself, it's that I didn't want anything to do with farmed chickens or the system that created them. I wanted to upgrade my diet to the most environmentally friendly foods I could get my hands on, which meant I had to start looking outside the grocery store.

To those who hate hunting and are vegan or vegetarian: You've made a conscious choice not to support the meat and dairy industry with your dollar. I'm glad you care about animals and the environment enough to change your habits for them. But you have to join me in the real world for just a moment and acknowledge that you aren't going to turn everyone vegetarian. Veganism and vegetarianism are not options for most people living outside of wealthy, industrialized urban areas. What you can do is be supportive of people who spend their money on more humane and environmentally friendly meat sources–and hunting in the United States is a part of that. If you want to help animals and the environments they thrive in, then hunters aren't your enemy. They should be your closest allies, because all they want is to expand habitats and make sure they're filled with wildlife. I would urge you to picket the grocery store meat aisle instead of the Fish and Game Department or your neighborhood's butcher. In the meantime, please stick around. Most people who find themselves opposed to hunting haven't had the opportunity to dive into wildlife economics or study how conservation funding structures work in the United States and abroad or have forgotten how focusing on individual animals can mean losing sight of larger environmental issues at stake–the ol' forest for the trees bit. We're going to look at both. Thanks for reading.

To those who are already hunters: You don't really need this book, but there's a reason you should read it. You represent hunting. That's a bigger deal than a tennis player representing tennis. If John McEnroe throws a fit and the public doesn't like it, they don't abolish all of tennis. But that is closer to the case with hunting. Hunters are the minority, and the public, regardless of how educated they are on the topic, are increasingly the ones making the rules about hunting.

Hunters like to talk about R3, the commitment to recruit, retain, and reactivate fellow hunters. But where is educate? It may not start with R, but it's just as vital. We'd have much higher attendance at weekend Learn How programs if Learn Why received more emphasis. How else to change the hearts and minds of people who may never even go hunting? It's not just new hunters that need to be brought into the community–it's new hunting allies. Frankly, they're even more important. Shrinking landscapes can only support so many hunters, but with more educated allies and everybody working together, efforts like habitat restoration could reach new heights of success. There's a massive untapped resource of potential advocates out there, people who love the outdoors and wildlife and want to see wildlands flourish. They might never take to the field, but they need to be included when we think about recruitment. A positive public perception might be worth more than the extra Pittman dollars when hunting issues end up in legislation, and it can only come from greater literacy about why people hunt and why hunting is valuable. As more issues are taken out of the hands of wildlife experts and handed over to John Q. Public, it behooves us to step up as educators. Adopt a non-hunter. The key talking points are the same ones that bring new hunters to the table: conservation funding, valuing the outdoors, habitat preservation, and environmentally sustainable (I argue beneficial) meat sourcing. Thanks for being so welcoming; let's keep it up.

At different points in my life, I've been a member of every group I just listed. I hated hunting as a kid and have pages in my childhood journals to prove it (reading them today, they sound like the furious ravings you'll find on social media). One year during archery practice at summer camp, I refused to shoot at a paper target with the outline of a deer on it. The counselor had to stop practice so he could walk out to where the targets were and flip mine over to a plain bullseye. I was a vegetarian for a brief amount of time—so brief it doesn't even count—imagine declaring you're a vegetarian after lunch, then taking it back before dinner. Though I continued to eat meat, I still hated hunting. I could not for the life of me understand why anyone would want to seek out animals just so they could kill them. I divided hunters into two camps in my mind: those who sought out trophy animals to display on their walls as part of some primitive machismo pissing match and those who hunted for "sport" and food, though I imagined the food part was just a bonus and a thinly veiled excuse to justify bloodlust. Why wouldn't you just buy your meat from the grocery store?

Thankfully, I've always been a curious person. This sometimes conflicts with my stubborn nature. Exposing yourself to new information is a great way to change your mind—but I hate changing my mind. Stubborn as I am though, I've always believed the adage that if you're the smartest one in the room, you're in the wrong room. This desire to constantly be learning new things has led me to a jack-of-all-trades lifestyle, which can be a blessing or a curse, depending on your attitude. But everything I've ever been curious about has always had a common thread: animals. I've followed them through art, history, economics, food, ecology, and philosophy. I never expected my obsession to foster a fascination and love of, first taxidermy, then hunting—though in retrospect it's not hard to see how all my passions would intersect there. Over the years, I became fascinated by how animals fit into the puzzle of our human-built environments, food systems, and our inability to extricate ourselves from the natural world. I learned that attempting to remove myself from nature was not always the best way to protect it.

If I had to boil down the themes of my personal journey to becoming a hunter, I'd name them value, visibility, and nuance. I'm hoping that because of who I am, a city-dwelling, animal-loving, left-of-center, locavore environmentalist, the messages I have might be more readily received by those who consider themselves similar. Though I hope the content of the message is the greater takeaway, I recognize that, for many, the messenger is often part of the message. Can we out-think our emotions or politics? I'm generally inclined to say no. But I was able to change my mind about something I never thought I could, let alone would. So, at the very least, I know it's possible.

1

The Butcher

Meat, Animals, and the Environment

When we arrived at the butcher, it was clear I wasn't the only one finding success in the early days of New Jersey's archery season. I filled out a form with my hunting license information and what I wanted done with the meat from my deer. It was like meat-Christmas. I asked for predominantly whole cuts and ground but couldn't stop myself from requesting some specialty items like kielbasa, landjager, and hot dogs (I really love sausages). Before leaving, I asked the butcher if I could keep my deer's head. I planned to clean and keep the skull as a memento of my first successful hunt. He strolled over to my deer and, with one swift and butter-soft knife motion at the base of the skull, removed the head. Luckily, I had another bag on me, and, after wrapping it up, I slipped my deer's head into my backpack next to his heart.

If you'd told me as a kid that I'd grow up to be a hunter, I would never have believed you. It would have been such a stretch of the imagination as to not warrant a second thought (though looking back now, I see all the signs were there). When I was a kid, all my opinions about hunting were shaped by my unconditional love of animals and my politician-level commitment to avoiding the investigation of something I was sure I was right about. I was so emotionally invested in animals that I was unable to, even momentarily, entertain the thought that there might be anything else at play in the mind of a hunter beyond a monstrous desire to kill.

Other kids, knowing my affinity for all creatures great and small, quickly found the best way to torment me was to torment animals. As my mother would say, I have a memory like a sieve, but I have vivid and haunting recollections of kids at camp trying to swat and kill all the dragonflies I was coaxing to land on me, an older boy taking the leopard frogs I had caught to show him and putting them in hard to escape places I couldn't reach to save them, or the time a few kids stomped on the head of a glass lizard I was following through the grass. Surely these horrid little sociopaths were tomorrow's hunters? Now I look back and see that childhood adoration for animals was what made me more likely to start hunting, not less.

When I used to think about my relationship to meat, I often felt like one of those drug addicts in recovery who'd visit our school and try to scare us kids straight. Only I'd never kicked it. Around the fifth grade, I truly faced the fact that the meat I was eating came from an animal that had to die because I wanted to eat it. I was practical about how I would handle this uncomfortable truth and simply told my parents that I was a vegetarian now and would no longer be eating any meat. They sort of shrugged their shoulders and dinner went on. Sitting at our kitchen banquette, I was left with the decidedly unsavory realization that my decision meant dinner now consisted of just potatoes and broccoli. I knew my parents would never adjust their cooking to suit my new dietary restrictions, and a cursory mental scroll of all my favorite things to eat revealed that every one of them was meat—hot dogs over cake 100 percent of the time. Or as Anthony Bourdain put it, "All my happiest moments seem to revolve around meat in tube form." My stint as a vegetarian lasted the pathetic minutes between my declaration and the point at which the smell of caramelized rotisserie chicken skin became too great for me. Perhaps six minutes if I'm being generous.

I did my best not to think about it. I felt bad about the animals but used my perfectly good kid-logic to temporarily assuage my guilt. Lions rip apart little helpless frightened baby antelope and I wasn't doing that, so that was something, right? But as I got older and learned more about meat, how it was produced and its broader relationship to the environment beyond the animals themselves, I became more conflicted.

Factory farms are so worried about people seeing inside their industrial meat monoliths, full of noisy uncomfortable animals all squashed together, that they have lobbied for and successfully passed "Ag-gag" laws at the state level, which prohibit the undercover filming of slaughterhouses and other agricultural related industries. They're afraid that if you see the reality of where your meat comes from, you'll be so disgusted and heartbroken you might never eat meat (or at least their meat) again. I like the idea that this might be true. I hope it is. I hope people would be so distressed over disturbing images of factory "farming" that they would give up their cheap, nondescript grocery store meat in favor of something better. But hasn't everyone heard the factory-farm horror stories at this point? Haven't there been enough "shocking reports, tonight at nine"? I've seen PETA hand out flyers with graphic slaughterhouse images on them to unphased New Yorkers in Union Square who take one glance before tossing them in the closest trash bin on their way into the market. (To be fair, New Yorkers might not be the best audience, as we're nearly impossible to unsettle. We have seen too much.)

When I was younger, the known and the unknown of large-scale meat-production industries made it difficult for me to reconcile my own meat-forward diet. I felt like I needed it more than I wanted it, and that was confusing. Vegetarians became a big turn on for me. I dated four of them (and to this day they represent a substantial portion of my friend group). I liked that they had made a conscious choice to not eat meat. They had thought about their eating habits and adjusted their diets accordingly. I practically considered them a more evolved people. The same way some folks are born without wisdom teeth, there were some people out there who were physically and mentally capable of being vegetarians–unlike myself, a hairy, club-dragging troglodyte. I think most people who eat meat don't think about it at all, it's just part of the meal, something on the menu. I've heard people who are far less meat obsessed than I am say things like "I can't stand vegetarians. Not eating meat is weird." And while it might very well be weird in our evolutionary history, I wondered why they cared so much. Had they missed all that media coverage of the environmental toll modern meat production was taking on the earth? I felt like we should all be kissing the asses of every vegetarian or vegan we met, for taking even the tiniest strain off the very planet we were all living on.

A 2017 study published in the journal *Science* calculated that, "eating no meat cuts an individual's carbon footprint by 820 kilograms of carbon dioxide (CO_2) each year, on average, about four times the reduction they'd get by recycling as much as possible. (Emissions generated by eating meat result, in large part, from the large amounts of energy needed to grow, harvest, and process feed crops.)"

I don't know what meat they were calculating. There are big differences between how cattle are raised and where they are coming from. A 100-percent-grass-fed steer that lives on a regenerative grazing ranch close to home is a world away from the environmental impact of cattle fed on lots that used to be rainforest and shipped overseas (the blanket term "recycling" is also unclear, as only 8–9 percent of plastic, 25–27 percent of glass, and 65 percent paper actually gets recycled). Still, as an average media consumer trying to make informed choices, I found myself inundated with reports on the environmental impacts of meat, from the land and water used to grow livestock feed to the land and water needed for the animals themselves.

The study in *Science* went on to reveal a more significant data point: "By choosing to have one fewer child in their family, a person would trim their carbon footprint by a whopping 58.6 metric tons–about the same emissions savings as having nearly 700 teenagers recycle as much as possible for the rest of their lives." In *The Climate Diet: 50 Simple Ways to Trim Your Carbon Footprint*, Paul Greenberg cites the amount as "sixteen tons of CO_2 per year for the rest of their life," for a kid born in the United States.

It made sense to me that fewer people meant less pressure on the planet and its resources, but I hadn't ever considered adding my child-free lifestyle to my arsenal of environmentally conscious actions, because I wasn't making that decision based solely on my concern for the planet. I just didn't want kids.

Tamar Haspel, *Washington Post* journalist and author of *To Boldly Grow* made my freewheeling lifestyle sound downright heroic when she wrote, "No amount of bean eating or Prius-driving will compensate for reproducing, and it's the childless, not the vegetarians, who are more likely to save the planet. Which doesn't mean that we should ignore the benefits of beans and Prii (plural of Prius) or that we shouldn't have kids–it just means that we should acknowledge that human survival takes a climatic toll. Our obligation isn't to minimize our carbon footprint at the expense of all other considerations; it's to try to be prudent, taking those considerations into account." Craig Chandler doubled down in 2019 for the Yale School of the Environment, saying, "No responsible series discussing finite global resources or long-term sustainability–and certainly not one on the challenges posed by a human-caused warming of the atmosphere–can ignore what many consider the best-left-unmentioned 'elephant in the room': global population. Simply put, it is for many an issue too sensitive to be raised, too divisive to be considered... but yet too important to be ignored."

Studies that quantify the environmental impact of certain behaviors rarely include the consequences of reproduction–tell people they should have one fewer kid for any reason (ecological collapse of our planet included), and they're liable to flip out. I find it strange when I hear proponents of veganism argue personal responsibility and sacrifice for the benefit of the earth only to discover they've produced a bunch of offspring. It's hard to take an impassioned message of "you must change this core behavior of yours for the good of our earth" from anyone who chooses to *have* multiple children rather than *adopt* multiple children. But the fact that people can't quite explain their reasons for "wanting their own" kid versus adopting one is enough of a statement about our limbic system lizard brain versus the rational prefrontal cortex we like to give all the behavioral credit to. We're animals, and animal behavior doesn't get more basic than wanting to eat meat and make babies.

Okay, no kids for me–that was my ecofriendly ace in the hole. I didn't want that to make me complacent about my eating habits though. This was about my relationship to animals, not people. And I knew my love of vegetarians (in or out of the bedroom) didn't count as an actionable measure or personal responsibility when it came to being a conscious consumer.

As I researched further, I realized that the alternatives that would have become my norm as a vegetarian didn't necessarily solve the issues of land use and animal deaths that concerned me either. In *The Omnivore's*

Dilemma, Michael Pollan says, "Killing animals is probably unavoidable no matter what we choose to eat." And "If our goal is to kill as few animals as possible people should probably try to eat the largest possible animal that can live on the least cultivated land: grass-finished steaks for everyone.... The vegan utopia would also condemn people in many parts of the country to importing all their food from distant places.... The world is full of places where the best, if not the only, way to obtain food from the land is by grazing (and hunting) animals on it–especially ruminants, which alone can transform grass into protein."

Industry and their brands try their best to make us feel good about their environmental impacts and our choices. Most are designed to make you feel better rather than do better. It reminds me of the 1990s, when "fat" became the bad guy. We were eating processed crap and the companies that made it didn't want to take the blame and change their business, so instead they leaned in, rolling out fat-free everything. In order to make up for the lost fat, they used cocktails of sugar to improve taste, making their products much unhealthier. Today, many of those same corporations have invested in "plant-based" products. Plants are much easier to manage than animals and are therefore preferred by suit-wearing conglomerates. New industry has been built within the economy of "greenwashing" in the same way the fat-free health movement was in the 90s. They're still nudging us toward processed junk food, but this time with greener marketing, feeding off our limited understanding of the natural world that's compounded by our peculiar human views toward other animals. No doubt we are headed for a similar reconning with processed plant-based foods. Imitation meat is not an environmental panacea, and it's certainly not a less harmful food choice than meat. Once we start shipping beans to and from energy-intensive processing factories so they can look and taste like chicken nuggets, we enter a state of diminishing environmental returns. If a concern for animals and habitat were the prompts for someone's vegetarian diet, then they should just eat the beans. As much as people might question the motives of hunters, I question the motives of some vegetarians. Why choose something meant to look and taste like meat if you don't want to eat meat? If your goal is a lower environmental impact, choosing the nugget because it's tastier than its vegetable source is hypocritical. And a goal of simply reducing farm animal slaughter, without considering the impacts on wild animals and their environments, seems like a narrow-minded motivation to me. A plant-based nugget may be marketed to those concerned about farm animal welfare, but this obscures the environmental impacts of the monoculture industry that produced the nugget, the wild animals displaced or killed in service to the land needed for its raw materials, and the factories needed to change it into its final nugget form.* In a piece for the Center for Humans and Nature, Mary Zeiss Stange, author, environmental activist, and conservation scholar, wrote, "By opting out of meat-eating, we cannot ignore the blood that is still, inevitably, on our collective hands. Mechanized farming is lethal to animals and their habitat, and a farmer harvesting a field of soybeans wreaks more carnage in a single sunny afternoon than your average hunter could accomplish in an entire lifetime."

These days, I like to compare the vegan "hot dogs" at the grocery store to the venison ones in my freezer. I'd pit my venison dogs against them in a battle of the environmentally, ethically, and nutritionally superior any day of the week. Tamar Haspel, author of the James Beard–winning *Washington*

---

*The pescatarian and casual fish eater can suffer from the same out-of-sight-out-of-mind philosophy as the grocery store patron buying cheap ground beef or a plant-based substitute, but to a potentially more environmentally calamitous degree. It's much easier to see drone footage of a forest cleared for cattle grazing or corn growing and clutch your pearls in horror, but all that fish stuff is happening underwater and far away from the eyes of landlubbers. If you want cheap fish for lunch, you're responsible for all the bycatch (unintentionally caught animals) that come up with the net. Unless you're fishing for yourself, you can't pick and choose. There are sustainable ways to eat seafood (bivalves baby!) but just like meat, if you're not going to catch it yourself, then it requires more work to seek out, more money to spend on responsible sources, and more homework to know what those responsible sources are.

Ret Talbot is a datacentric writer and photographer who's spent over twenty years reporting on marine ecosystems and their intersections with the seafood and aquarium industry. His wife Karen is a talented scientific illustrator and wildlife artist—the majority of her subjects are fish too. They live on the coast of Maine with a bunch of entertaining ducks and eat seafood around four times per week. They're an oceanic power couple, tireless educators, and evangelists for our aquatic friends.

When prompted, Talbot's succinct but difficult-to-execute answer to sustainable seafood is "know your fisherman." The same sentiment is applicable to farmers on land, and he recognizes this is easier said than done but also emphasizes that that's the point. Maybe it

Post column "Unearthed," had a similar revelation when she wrote about her experience crafting her book *To Boldly Grow*, "I spent eight months writing a book about the good things that happen when you put down your phone, roll up your sleeves and go outside to find something to eat, so I naturally also gave some thought to the environmental implications of those foods. Turns out, one of them absolutely tops the environmental charts. It's unequivocally the single most ecologically friendly food you can eat. A food that actually makes the environment better rather than worse. Seriously. Literally. The food is venison. The catch, of course, is that you have to kill a deer."

For years, I existed as a guilty meat eater, thinking it could only be done one way, and that way was bad for the earth. So, if I wasn't going to stop eating meat, then I wanted to do all in my power to eat as responsibly and ethically as I could. Once I started to earn enough money, I was able to buy the more expensive meat that came with labels like organic, grass fed, local, certified humane, and animal welfare approved. The changes in our food system that are needed the most to combat the climate emergency and animal cruelty really have to come from larger government regulations, but even if it was just a drop in the bucket, and even though I

---

should be more difficult or more expensive to buy swordfish period, but it should be especially difficult in a place like Colorado. It's the same message I push when it comes to meat—but even I have trouble with seafood labels sometimes. Talbot says to buy "US-produced seafood and stick to state waters over federal waters if you can." He was quick to say how that can be a tricky prospect when so much seafood is caught in the United States but then shipped out to China for processing then shipped back to the United States for sale. When I asked him about seafood pricing, the answers continued to mirror the issues we see in terrestrial meats. "People are so used to food being cheap." None of us *wants* to spend an arm and a leg for a filet or a ham, but the artificially low prices of meat and seafood only perpetuate the most damaging and unsustainable aspects of these industries. Talbot advises to "buy local and buy in season. That goes for any food system."

But beyond knowing your fisherman or buying local, my biggest takeaway from Talbot was "define what sustainability is." Thinking about my own definition, I realized it had a lot to do with where money was going and how populations and environments would rebound year after year. But there was always some outlying factor that would lead me to boiling down my definition to that of Supreme Court Justice Potter Stuart's—I know it when I see it. Short of inspecting every fishery on earth personally, it behooves the conscientious seafood lover to put a little effort into the food they buy, both for their own health and the health of the sea. The ocean is not bottomless.

Venison Hot Dogs TK

Caption TK

wasn't sure if I could trust the labels, I felt good about voting with my dollar. My newfound pride in purchasing the most environmentally friendly meat I could afford coincided nicely with the farm-to-table movement hitting the mainstream. Suddenly people seemed to care about where their meat came from. They wanted heritage breeds of animals from local farms they could theoretically drive to. Farmers markets became chic, and having the name of the farm your pre-pork-chop pig came from was the latest trend on menus. Restaurants started to offer offal. I was giddy when I saw beef heart on the menu at a restaurant in Chicago. As better and more diverse meat became readily available, I started to feel a little less guilty when I bought it.

There's a good deal of myth surrounding the birth of the farm-to-table movement in America. A generally agreed upon milestone, however, is the 1971 opening of Chez Panisse in Berkley, California, which is regarded as the first farm-to-table restaurant in the United States. According to the Farmers Market Coalition, in 1994, the USDA Farmers Market Directory had under 2,000 listings of registered markets—that number was over 8,600 in 2022. Michael Pollan wrote *The Omnivore's Dilemma* in 2006, and it continues to be the apogee of revolutionary food writing for a generation. My personal meat revolution started when Camas Davis started the Portland Meat Collective in 2009, a program focused on teaching whole-animal butchery classes and educating students (including kids) on what responsible meat consumption looks like. I lived in New York at the time, but the story of the writer turned butcher who was teaching butchery classes to anyone who wanted to learn spread to meat connoisseurs and conscious eaters quickly. Here was someone who not only believed people wanted more transparency in meat production but also the opportunity to have a hand in it themselves (the Good Meat Project, the nationwide program The Portland Meat Collective gave rise to, is still going strong today). I was eager to learn whole-animal butchery, but classes like those at PMC didn't exist in New York yet, not that I could find. So, I did what had been successful for me in the past, door-to-door cold calling. I walked into butcher shops that I liked but didn't frequent and asked if they'd take an apprentice. Most of these places were run by old men who knew better than to waste their time with a slow-learning novice, but I found one place so new that the appeal of free labor must have clouded their better judgment.

Ben Turley was one of the two owners of The Meat Hook butcher shop in Brooklyn (he left the shop in 2022 for new meaty pursuits). He is also one of the kindest, most generous people I've ever met. The moment I set foot inside the shop, I was delighted by what a bright, spotless, shrine to meat it was. I felt like Charlie meeting Willy Wonka and immediately asked if I could apprentice with him. My interview went like this:

Ben: Have you ever worked in a butcher shop?

Me: No.

Ben: Have you ever worked in a kitchen?

Me: No.

Ben: Have you ever worked in food services at all?

Me: No.

Ben:

Me:

Ben: Okay. Can you be here Monday?

I had no illusions about disassembling a pig myself for the foreseeable future and was perfectly happy to do all the lowest jobs on the totem pole while I learned how the shop ran and how the literal sausage got made. To that end, I started with untangling bags and bags of pig guts. Imagine twenty pairs of corded ear buds are thrown into a plastic bag full of thick brine and tossed around by a paint mixer, but instead of ear buds they're pig intestines. I would open the bags and separate each length of pig gut, wrapping it around the widest part of my hand. Seven or so inches before reaching the end, I'd slip them off my palm and use the last dangling bit to wrap up the center and make a tidy little pig-gut bow tie. This made it easy to grab a length for the next batch of sausages. It was a cathartic process, and I enjoyed seeing myself get faster and faster at it–pretty soon I was blowing through two bags in the time it used to take me to do one.

Look at me, bragging about my pig-gut untangling prowess; apologies.

The shop received beef, pork, lamb, and chicken. And the way everything looked when it came into the shop gave me the reminders I had been searching for: that meat did, in fact, come from animals; it wasn't just amorphous plastic-wrapped chunks–it was part of a once-living body.

Pig day was always especially meaningful to me. The pigs hang in the delivery truck and are split nose to tail. One by one we (the four to five people working that day) would line up at the truck and each take a cold side of swine on our shoulder into the shop and pile it on the main cutting table. They're very heavy, and between the weight and the awkward angle you have to carry them, my back was usually spent after three. They looked like pigs too. On the one side, they had skin and a face and one of those wonderfully spade-shaped ears, and on the other, they were a beautiful, pink, 3D anatomical specimen. Pig and pork in one.

Most of the butchers at the Meat Hook said they didn't eat much meat themselves because they had become so finicky about its sourcing. I began to hope that this ceaseless parade of animals coming in one door and meat leaving the other would curb my own desire to eat meat, perhaps turn me off it completely, but it didn't. I just became more dug into my personal quest for meat perfection. I was having a real hand in the production of my meat, and when I bought meat, I knew how it had gotten to me–more than any meat I'd ever eaten prior. I didn't even have to purchase much at all, as meat was the currency with which I was compensated for my work (a deal you should be suspect of unless you're working at a butcher shop).

Another job reserved for me was carrying the five-gallon chore bucket full of pigs' blood from the delivery truck to the meat locker. The sides of the bucket were often wet, and blood would smudge the bottom half of my apron and drip onto the toes of my topsiders. One day I was carrying the blood bucket in after taking three pigs off the truck, and the top wasn't affixed all that well. Blood sloshed from side to side, spilling out from under the lid. I tried to carry it gingerly, but the bucket seemed to get heavier and heavier, and, holding it with two hands, I couldn't help but swing it a little. Blood is not only thicker than water in a metaphorical sense–it also weighs more. For some reason, I chose to walk through the picturesque event room that was used for cooking classes, instead of through the main grocery door.

A few little stumbles caused by the bucket smacking into my shins were enough to ensure a thick steady stream of blood trailed me through the event space, into the butcher shop and meat locker. After putting the bucket away, I rushed to clean up the mess but wasn't entirely sure how to begin. I started to use paper towels, hoping to sop up as much liquid as possible. Then I brought out the mop. This led to smearing blood all over the floor of the Instagram-worthy event room, making it look more like a David Cronenberg film than a Nancy Myers one. Ben seemed unfazed by my melee with the bloody mop and newly red floor. Eventually I cleaned up the mess, and the incident was never spoken of. The Meat Hook offers their own butchery classes now—no more random recruits.

Kris De la Torre, a sustainability fellow at the Harvard Graduate School of Education and an expert in the intersection of food, people, and the environment, describes the industrialized food system "as a mechanism for obscuring relationships." She declares it criminal that "we get away with paying so little for farm or ranch-raised meat," and explains, "after working on a few different livestock operations [including Dan Barber's two-Michelin-starred Blue Hill at Stone Barns, chosen as one of the World's 50 Best Restaurants in 2015's industry-coveted William Reed List], it isn't only the physical labor of feeding, cleaning, and moving animals that overwhelms me. It's that the average consumer has no idea, could not even imagine, the dedication it takes to raise animals." In her experience, "the act of harvesting farm animals starts with the intentional care you give them for months ahead of the slaughter. Knowing all along that these animals are being raised, essentially in service of feeding people, made me acutely aware of how I took care of them and the kind of life they were able to lead up to the moment of harvest."*

---

*This accountability is a theme in Kris's farm-based work. She describes harvesting chickens, turkeys, and rabbits as "an intimate experience," explaining, "the tools involved in slaughtering small, farm-raised animals significantly increase your proximity to the animal. The operations where I was tasked with harvesting birds and rabbits were small enough that the tools on hand mainly consisted of a sharp knife, in some cases a cone in which to place the bird, and the assistance of another farmer who would hold the animal. I always felt an extreme responsibility not to fuck it up, to cause the least amount of pain, in the shortest amount of time.

"What I loved about working with livestock is how attuned you become to their cues. It astounded me how observable it was that my energy affected the animals, even just during

The further we get from the farm, the water, or the woods, the more foreign our connection to food becomes. In her nature-based education work, Kris cares deeply about and tries to emphasize "the importance of fostering a personal connection to the outdoors, both wild and cultivated," acknowledging that for many people, it's hard to "know the difference [between wild and cultivated areas] when it all feels abstract, inaccessible, or even scary." Visiting a butcher shop (rather than the prepackaged meat aisle) might require more comfort with larger cuts of meat or knowledge of animal anatomy than the average urban or suburbanite possesses. People who don't know how to ask for or cook less-common cuts of meat might not be able to make the connection between animals and food.

The language we use to talk about meat can also create distance. When I saw that beef heart on the menu in Chicago, I wanted to order it but was asked not to by my vegetarian then-girlfriend—heart was too recognizable in name, with too many poetic connotations. Words like "New York strip" or "chop," were vague enough for her to psychologically stomach sitting across from me while I ate. Plenty of meat-eaters hide behind this kind of language too. Divorcing the meat we eat from the live animals it comes from makes it easier to see hunting as further from food rather than closer to it, but in reality, there is no closer relationship to meat than what the hunter pulls from the woods or the angler reels in from the water. You have to be on the land to hunt it, you have to see the animal alive in front of you before you can kill it, and you have to take the meat from the bone to eat it. If you make yourself a part of this cycle, then each aspect of it becomes personal. And we care about what's personal to us.

---

feeding time. During a harvest, I was aware that every move I made had a direct impact on the animal's (hopefully brief) experience of what was happening to them. I always felt accountable to the animal, down to the pace of my breath and the weight and steadiness of my hands. I don't think this is something specific to who I am. I worked with high school students two seasons in a row on a farm where I was able to take them through the rabbit harvest. Most participated by holding a rabbit or using a harvest knife to slice the throat and spinal cord of the rabbit. During both seasons, young people who struggled to demonstrate even the smallest amount of self-control instinctively approached our work with composure they hadn't shown at any other point in the trip."

While I was working at the Meat Hook, I became a regular at my favorite restaurant. It was a short walk from the shop and on the way to my train home. This cozy haunt has a short chef's counter, which is the best seating option when dining alone or with one other person. Perched up on the barstool, I could watch the three chefs deftly navigate their minuscule cooking space and savor all the sights, sounds, and smells of ingredients becoming meals. One night, a couple on a date (let's call them Brad and Cora) was sitting at the only two-top, next to me beyond the end of the chef's counter. I'm not entirely sure how many dates in they were—more than one, less than four—and their giggling flirtations were tough to ignore, so I looked over occasionally to spy on what drinks and appetizers they were ordering. Then their entrees came out. Everything is à la carte at this place and is best when shared, so you can try more things on the menu, but the steak went down in front of Brad and the chicken was put in front of Cora.

There are a few steaks on the menu at this place, and while they're all good, the hanger steak is a fan favorite—first, because of its reasonable price (considering its source and trendy locale), and second, because hanger steak is a damn fine cut of meat that most people don't even know exists. There's only one on every cow, and it gets its name because it hangs from the diaphragm along the lower part of the belly. Back in the day, it was just easier for a butcher to take it home for himself than bother to explain what the hell it was. Hence the oft-used nickname "butcher's steak." No trouble there, Brad was foaming at the mouth ready to cut into the salty, buttery, medium-rare delight before him. But Cora was visibly upset.

You see, the chicken at this friendly little restaurant, with the names of all the local farmers they work with scribbled out on the blackboard, is a whole chicken. And perhaps when I say "whole chicken" you think of what a rotisserie chicken looks like. But at this restaurant they actually mean whole. It has a neck, a head, legs, and feet. Cora was positivity beside herself. Her mouth went from partially agape with shock to clenched shut in abject horror and disgust. She couldn't even look at it let alone start carving it up. Brad looked on, not knowing what to say or do. They were both quiet. And he didn't touch his steak in solidarity.

I went from feeling a flash of bemusement to full-on, red-faced fury. I wanted so badly to get up from my stool, pick up her chicken, hold its charred little head right in front of her face, move its roasted feet around her plate like a marionette, and say, "This is a chicken. This is what a chicken looks like. When you eat chicken, the meat comes from a little bird that looks just like this one. A couple days ago this bird was walking around a farm, and now it's on your plate. That's how meat works. Is this news to you? Were you unaware that animals become meat? Or did you just never ever want to be reminded of it? If you can't even *look* at this chicken, then you sure as hell shouldn't eat it. Nor should you ever be allowed to eat any meat ever again!" Instead of this fantastical self-righteous outburst, I waited for the waitress to top off my water and suggested she might portion the chicken out before Cora had a stroke.

I think of those two a lot. And how important that chicken's head was. It doesn't matter to the chicken if it's served with or without its head. It doesn't matter to the buffalo if you use every part of it. It's dead. It doesn't care. But what does matter is how your acknowledgement of the animal's life informs the way of its death. We've removed ourselves from the environments our food comes from (vegetables included) to such a degree that people really do seem think it all just comes from the grocery store. If you don't see your meat as former animals, then why would you care about how they lived or died? Acknowledging they were animals seems so much more productive to me than just respecting them postmortem. I hate terms and phrases like "respect its sacrifice" and "gave its life." I promise you the animal did not *give* its life to you. A steer did not walk onto your grill and say, "I understand you need burgers for the Fourth of July weekend, I volunteer as tribute."

I'm still not a perfect meat eater. I'll get an In-N-Out burger if I'm in California, and I'll order meat dishes at restaurants even if they don't provide a full "Colin the Chicken" style dossier on the provenance of said bird (see *Portlandia* season 1, episode 1). But I prefer and strive to eat at places that do. These days, I look back on my vegetarian exes and realize it wasn't their diets that was doing it for me, it was that they were thoughtful people. Today, I see the same care, thoughtfulness, and consideration in my hunting pals. People who are adamant about getting meat right—and for them that means wild habitats, natural diets, and a fair chance at escaping the dinner table.

The Omnivore's Dilemma brought the farm-to-table movement to the mainstream. It might be responsible for the earliest iterations of the field-to-table movement as well. You didn't have to eat at a specialty restaurant, or take a cooking class, or live near a farmers market, you could just bring that 450-page book home and learn about where your food was coming from and how you fit into the food chain. Because, regardless of what you eat, you are a part of the food chain (I think that came as a shock to some readers). There's one chapter about eating animals in general (Chapter 17) and one on hunting (Chapter 18), and in those fifty-nine pages, Pollan eloquently sums up what so many books (including this one) take cover to cover to analyze. (Talented bastard.) Nearly two decades later, it's still a must read for the curious and concerned eater.

With the publication of *Omnivore*, people began to examine what we eat under the stronger lens of food's broader implications for the earth, not just one's health. Shortly after, a flood of new works hit the market, specifically about starting to hunt for your own food as a rebellion from the ecologically destructive and cruel conventions that made up your average store-bought meat. In 2011, there was *Girl Hunter* by chef Georgia Pellegrini, *Hunt, Gather, Cook* by chef Hank Shaw, and *The Beginner's Guide to Hunting Deer for Food* by Jackson Landers. All three included recipes. The year after that saw publication of *The Mindful Carnivore* by Tovar Cerulli, and maybe my favorite of the new hunter series, *Call of the Mild* by Lily Raff McCaulou. Tim Ferriss, known for *The 4-Hour Workweek* and *The 4-Hour Body*, discusses his turn to hunting after a childhood of despising the slob hunters he saw while growing up on Long Island, in his 2012 book *The 4-Hour Chef*. It includes six pages of fifty-one color photos illustrating Tim's journey from his first grip 'n grin photo, through the process of gutting and butchering, to cooking and wrapping meat for the freezer, including his final presentation of a handsome doe-skull "trophy," the only physical memento that remains of that experience now the meat has long since been eaten. *Nature Wars* by Jim Sterba was also published in 2012, but it's less food specific and more about the complexity of modern wildlife management strategy and conservation among growing human populations.

Emma Marris (whose work has been compared to that of Rachel Carson and Aldo Leopold) writes about the changing relationships between people, land, and animals. She saw these hunting related books fill the shelves and, in a 2012 article for *Slate*, listed "the evolution of the new lefty urban hunter" with surgical accuracy, laying out the progression like this:

2006: Reads Michael Pollan's *The Omnivore's Dilemma*, about the ickyness of the industrial food complex. Starts shopping at a farmers market.
2008: Puts in own vegetable garden. Tries to go vegetarian but falls off the wagon.
2009: Decides to only eat "happy meat" that has been treated humanely.
2010: Gets a chicken coop and a flock of chickens.
2011: Dabbles in backyard butchery of chickens. Reads that Facebook founder Mark Zuckerberg decided to only eat meat he killed himself for a year.
2012: Gets a hunting permit.

I live in an apartment, so I've never had a vegetable garden or chickens, but the rest of Marris's list checks out (for me as well as many other hunters I know). My half-assed stint as a vegetarian, the commitment to only buy "happy meat," showering myself in books on the topics of conscious eating and sustainable foods, going to farmers markets, working at a butcher shop, learning to hunt–it's eerily accurate.

I spoke with Marris about what changes she might make to her list for its ten-year anniversary. She didn't have any sizable edits–perhaps the addition of an attempt at eating less meat overall and the occasional sampling or adoption of "imitation meat"–and neither did I (thankfully, since me "editing" Emma Marris' writing would be like my fifth-grade soccer team critiquing Pele's footwork). She was, however, a little dispirited that after all the writing on the subject and the momentum it generated, the farm-to-table movement still hadn't transitioned to a more mainstream awareness and embrace of the field-to-table movement. I'm more optimistic (maybe because I've only spent half as much time dwelling on the topic). If those books and articles represented the trend pieces of the day, then maybe we are now, ten years later, ready to see the ideas they fostered become more widespread.

For myself, I was proud I was working at the butcher shop and proud I could look my pork and chicken in the face, but I knew I was still a step removed from the process of an animal becoming meat. The unfortunate middle business of killing and dying was something I'd have to take part in if I really wanted to practice what I was so sanctimoniously preaching: If you can't kill it, then you shouldn't eat it. I firmly believed that. I had outlandish fantasies of implementing meat-eating licenses that would require anyone desiring to eat or buy chicken, pork, or beef to first kill or bear witness to the death of one chicken, pig, or cow personally. My environmental, animal-centric tyrannical rule would unfortunately render my friend Jenn, who loves to cook and cannot stomach the sight of blood, a vegetarian. But then... she was more than excited to cook the venison roast I gifted to her and her husband Steve–maybe I'd add a game meat addendum? I admit I have not fully explored the logistics of my dictatorship.

Sophie Egan, an author whose work focuses on food's intersection with personal and planetary health, knows what it means to eat sustainably these days: the misconceptions, the trends, and the sneaky way the food industry can take advantage of those things. She spent five years as editorial director and director of health and sustainability leadership for the Strategic Initiatives Group at The Culinary Institute of America, and she's a food writer who has made her career studying and reporting on almost every aspect of the industry for just about any publication you could name. Her first book, *Devoured*, is a bit of a food-related *Freakonomics*, and, in 2020, she published her second book, *How to Be a Conscious Eater*, a thoughtful guide that tackles subjects like understanding food labels (which can be bizarrely complicated), how to cut back on plastic, knowing the difference between types of fats, and how to eat meat and seafood with a more critical eye for nature's benefit. She knows her shit.

We both tend to break food down into three categories: 1) is it good for the earth, 2) is it good for me, 3) is it good for (I say) animals or (she says) others (while my focus was always animal and environmental welfare,

Nature
Makes a Deer

A lady buys a hunting license and a box of rifle cartridges.

$ from the license and 11% of the cartridge costs go back to **US Fish & Wildlife**.

She shoots the deer on **Public Land** that's maintained with the $ made from the purchase of her license and rifle rounds.

The lady has **sustainable**, eco-positive venison for her Taco Tuesday party, an appreciation for where her food comes from, a connection to nature, some exercise, and a citizen scientist degree in ecology and biology.

An **Industrial Monoculture Farm** uses massive amounts of land and water to grow crops for factory farmed animals and plant-based meat substitutes.

The **Factory Livestock Farm** raises those animals in cramped, inhumane, unsanitary conditions. The **Industrial Monoculture Farm** depletes the soil of all nutrients eventually leading to soil death.

A **Processing Plant** is needed to turn animals into meat (a frightening environment for the animals) and monoculture crops into imitation meat.

The deforestation necessary to build and run the industrial agriculture farms, factory animal farms, and processing plants leads to accelerated **Climate Change**, and their byproducts pollute our air and water.

A guy buys some ground beef and imitation meat at the market for his Taco Tuesday party. He has successfully ignored his contributions to **global decline**.

FPO

Egan reminds me that food is not just the strawberry or the chicken breast but also the many people responsible for getting them to you). We also both agree that "plant-based foods are not automatically healthier," and that "being a conscious eater relies on a nuanced assessment of what you're putting in your body." And perhaps most importantly, that the cost of meat in this country is too artificially low and that it would be better for people, animals, and the environment if buying meat reflected the true cost of its production.

In prepping for this book, I wondered if Egan had ever given any thought to the health and environmental benefits of hunting in her food-centric career, and she was kind enough to chat with me about it. About 87 percent of Americans support hunting for food when questioned, but how many actually consider it a source of food without being prompted? Egan was gracious enough to tell me, "Hunting hadn't ever really crossed my mind before, and I write about this stuff for a living."

She had, however, broached the topic once with her cousin Riley in Montana, who's an avid hunter. Riley hunts deer, elk, bear, and birds and does all his own butchering and processing. "He hasn't bought beef or chicken in at least four years." Listening to him, she was in awe of how he hunts with a bow and spends months carefully planning for the upcoming season, including taking long hikes into the mountains for on-foot scouting. Then, if he's successful, how he has to "pack over 100 pounds of meat out of the backcountry on foot, amid potential predators like wolves and grizzlies."

When asked about the ethics of hunting, Riley echoed a lot of hunters' feelings about the hypocrisy inherent in that question. "People who are like, 'How can you do that? Hunting is cruel' and that night are having a burger for dinner–all you've done is paid several people to pay other people to do that [kill and prepare your meat]. And you don't realize you're eating a collection of twenty different lives." He's referencing the fact that ground beef burger patties are made up of meat from multiple cattle, not just one–but 20 is an absurd number. Absurdly low. Some estimates say the average burger patty from a large-scale producer can have bits of as many as 100 or more cattle in it. But not Riley's elk burger–there's just one elk in there, and he knows exactly where it came from. Sophie was impressed, calling this the "ultimate transparency." There's no better way to know where your food came from than getting it right out of the woods (or river, or sea, or sky) yourself. That plus butchering it on your own means the meat has only ever passed through one set of hands, yours. It doesn't get more transparent than that.

All the research I was doing in my quest to be a more responsible and environmentally friendly meat eater kept leading me toward hunting. The clues I found were big ones—whole loaves of bread, not crumbs. I began to change my mind about it the more I learned and had to consider, which made me realize that my blind emotional devotion to my childhood anti-hunting views had been the death of investigation into my adulthood. Science, which I've always looked to as the only impartial judge, will tell you, "Yeah we have an opinion, but we can change that opinion based on new information." Ethics, I believed, were best served when they contained a base of fact and reason, with only a finishing sprinkle of one's personal mores. What's funny is that if we were discussing, say, best medical practices instead of animals, I would have applied these principles without a second thought. All my other strong opinions (and there are many) were based on a near-robotic level of devotion to facts, logic, and reason, but when it came to animals, emotion ruled. As I started to change my mind about hunting, I worried that I had let my opinions on the one thing I cared about most be shaped by uneducated, emotionally driven feelings instead of sound scientific education and reason. I could tell you every fact about bears but nothing about the non-environmental systems at play needed to keep them on this earth into the future. I felt like a doctor who had been operating on family members this whole time.

While I had fond memories of catching trout in the Pocono mountains with my dad and trapping crabs for my mom to boil in South Carolina, that was the extent of my own personal experience dispatching creatures so I could eat them. I figured my first hunt would be the last test. Either I would be able to kill something myself and eat it, or I'd be so distraught from the event I'd never *want* to eat meat again, and I was okay with that as a possible outcome. It was about time I learned to hunt.

2

The Outdoorsman

Hunting Is Conservation

It was a wonky nervous smile that managed to squeeze out of my still very much shocked face when I heard the camera click, but I was proud. Not proud that I had killed a deer but proud of what killing the deer meant. I had been patient. I had made a good shot. I was finally responsible for the meat going into my freezer. And the money I spent to make it happen was headed back to the woods and the animals, not the industrialized agricultural system. I had achieved a goal that took me years of work and naval gazing to reach.

If you want a driver's license, you have to study the rules of the road then take a driving exam before the state will issue you a license. Same goes for hunting. You study an online curriculum, passing tests at the end of each section before you can move onto the next one, until you reach the end and receive a code that verifies you completed the whole course. Sometimes, if online study isn't an option, you can get a practice book from your local Fish and Wildlife department. Then you must attend three, three-hour in-person classes or one nine-hour class (because of the Covid-19 pandemic, most of Hunter's Ed moved entirely online, and I wouldn't be surprised if many of the classes stay that way into the future). These classes will review the whole of the curriculum you just finished. Then you take the final exam.

If you pass, you're issued your Hunter Education Certificate, and *then* you can register for a hunting license.

Most states now have an apprentice hunter license as well, which allows you to go hunting without having taken Hunter's Ed as long as you're with a licensed hunter. You will still have to take Hunter's Ed eventually, but being able to "try before you buy," so to speak, is an effective way to expose newcomers to hunting and get them excited about it without requiring them to commit to classes just yet. Hunter's Ed does not teach you how to hunt. It does give you some basics, but it's focused mostly on understanding the safety aspects of hunting. It also touches briefly on the government systems that control and fund hunting and conservation in America.

My Hunter's Ed class was in Queens, at a little shooting range I didn't even know existed. There was a single-day nine-hour class at a sporting club much closer to me in Brooklyn, but all the spots were full by the time I went to sign up. I couldn't believe it. I was shocked that multiple Hunter's Ed courses would sell out in New York City. I also couldn't believe I now had to go to Queens three nights in a row, a waking nightmare for anyone living in Brooklyn. My instructor was straight out of central casting, a crotchety old Long Islander who said "cawfee" with a thick New York accent. That accent was declared practically extinct in 2013, as reported by the *New York Times*, and I had never heard it in all my years in the city–I thought it was the stuff of linguistic myth. The two owners and managers of the range were also three-camera-sitcom characters–at least once a class, they found a way to insult New Jersey. It was surreal, to say the least. As far as the other students went, I was surprised at how diverse our class was. Like a microcosm of the city itself, the group represented a broad range of age, race, and gender. Far more diverse than I was expecting.

I liked the course; it was fun to be back in school for something I was interested in. One highlight was when we received a visit from two New York State environmental conservation officers (ECOs)–like when firefighters visit your school to teach fire safety. I was in my thirties by then but felt like a little kid, in my plastic chair with my workbook on my lap, staring up at the hunky officers in their smart green uniforms and wide-brimmed campaign hats that were pitched down to their eyebrows.

There are lots of wildlife-related jobs I will always lament never having. I wanted to be the next Steve Irwin or study great white sharks off the coast of California or trap and tag bears around the world. When, on career day in first grade, kids stood one by one at the front of the classroom to talk about wanting to be astronauts, firefighters, and doctors, I leapt at the opportunity to share the details of what it takes to be a game warden in southern Africa. I brought a photo I had torn out of my National Geographic of a man, armed to the teeth, standing next to a grazing rhinoceros, one hand casually resting on the rhino's back while the other cradled a rifle that must have been forty years old by the late 1980s when the ad first appeared. I couldn't think of anything better than hanging out with rhinos all day and defending them from poachers. It still sounds good to me, despite how dangerous it is. (The legal, regulated hunting of rhinos and other big game is most often what pays for the rangers who protect them from the indiscriminate hands of poachers, but the complexities of such a counterintuitive cycle would have been difficult for my first-grade self to understand.)

Palm Beach County Florida in the early 1990s made for a childhood that was equal parts *Tom Sawyer* and *The Philadelphia Story*. One day I'd be Easter egg hunting at Lilly Pulitzer's house, and the next I'd be catching green anole lizards and holding them next to my head so they could bite my earlobes and create the finest all-natural swamp-child accessories. I'd come home in the afternoon wearing my private school uniform and head straight for the woods only to race back to the house so my mom could hose all the newly gathered fire ants off me. Where most kids got snow days, we got cottonmouth days, in which parts of the school would close so wildlife relocators could sweep the buildings for venomous snakes (I pestered them relentlessly about their jobs). While playing on a neighbor's docked boat, I saw a manatee and her little calf munching on vegetation—from that day on I'd walk the canals as often as I could to find more.

I loved learning about animals and their habitats in *Zoobooks*, *National Geographic*, and countless field guides, but the stories I gravitated to the most were the ones that put me in those habitats with them. My family wasn't particularly outdoorsy, and since I couldn't be an animal myself, or be among them as I might've wanted, I liked to read stories about living with them in the wild. My favorites were spooky tales of wilderness survival, anything that would combine animals, adventure, and a surplus of danger, hence my attraction to the perilous jobs of game warden and conservation officer. Books like *Hatchet* by Gary Paulsen, a perennial favorite of restless outdoor kids, and anything by Jack London or about polar exploration were early favorites. I adored sharks and returned to a book with a vivid description of a shark attack in Australia the most. I watched *Jaws* on repeat—fancying myself a young Matt Hooper (the film's marine biologist), I'd pretend the bed in our guest room was a raft—not afraid of the shark but thrilled by it (a slightly different response from that of most people, which kicked off an unfortunate shark-killing spree following the film's release).* While at my dad's friend Richard's house in Red Hook, New York, I came across a paperback copy of Jim Corbett's *Man-Eaters of Kumaon* in his library. I liked the picture of the tiger on the front and the title had instant appeal. Inside was Corbett's firsthand account of hunting "man-eating" tigers across India in the early 1900s, the most famous of which had killed over 436 people. (Corbett's stories are incredibly brief. If you want more color on the whole crazy drama, then pick up *No Beast so Fierce* by Dane Huckelbridge after.) When I finished, Richard recommended *The Man-Eaters of Tsavo* to me, so I traded Indian tigers for African lions and kept reading, hoping I'd find *Man-Eaters of the Hudson Valley* next. My takeaway from these stories was that I liked when animals got the chance to eat some people for a change. It would be years before I could take in the depressing reality that human-animal conflicts like these are always the result of an animal's injury, old age, mistaken identity, or the loss of their prey and habitat—"man-eaters" aren't really a thing.

As I grew up, I took every opportunity to work with animals and in the outdoors. My first real job in high school was at the South Carolina Aquarium in Charleston, where I got to care for the alligators. I wasn't too concerned about being eaten by them, as they were only ten inches long. Then, in college, after falling in love with white water, I got my raft guide

---

*Jaws is still my favorite movie. The first movie I ever saw in a theater was the 1990 creature feature *Arachnophobia*—as such, it's a wonder I'm not afraid of spiders, but I suppose no surprise I find invasive species so fascinating.

license in Wyoming on the Snake River. My sister, successful with her own business already, was generous enough to pay for my residency at a Super 8 Motel while I worked on the river. I loved being surrounded by other weirdos who enjoyed getting pelted with hail in the wee hours of the early spring high-water season. When we opened for customers, the questions they'd ask me were often a reminder that the average person wasn't as environmentally literate as my colleagues and friends in the outdoor industry. A query from one customer–"At what elevation does a deer become an elk?"–confuses me to this day. After college, I got a job as a kayaking instructor and guide on the wild and wooly Chicago River. In fact, I spent so much time on the water that the University of Illinois used me as a lab rat one summer to monitor the river's water quality. They paid me eighty dollars per "waste material sample" I gave them. Not a bad gig.

When I first moved to New York, I worked as an outfitter for the outdoor goods companies Eastern Mountain Sports (now closed in New York City) and Recreational Equipment Inc. (better known as REI) so I could continue to build up my stores of gear. Quality camping and outdoor kit is notoriously expensive, and I could only afford it with employee discounts and industry pro-deals. I truly enjoyed my time at those places. It was fun to help people gear up for their trips, and, being in New York City, we were often outfitting wealthy customers on their way someplace exciting. I built countless kits for Kilimanjaro and backpacking trips around southeast Asia and one for a young woman who just wanted to hang out at Everest basecamp for a couple weeks with no plans to climb beyond that. Mothers would come into the shop in spring, preparing their teens for elite summer camps. More than once, someone just handed me their credit card and a gear list provided by the camp. It was hard not to go hog-wild. I had to remind myself it was their shopping spree, not mine. I helped a handful of people build "go bags" from the bag up. One guy, after some gentle taunting from me that a go bag is only good if it's near you when you have to go, promptly told me to double everything. He's figuratively loaded for bear now, so long as disaster doesn't strike while he's on his way to or from his Midtown office.

Some people would come into the shop and tell me they were just getting into camping and needed help picking out everything and lessons on how to use it all. I always started by telling them I didn't work on commission. That was my favorite maneuver for inspiring trust in the equipment I'd pick out for them—they'd know I was focused more on choosing the best gear for them, not the most expensive. Ironically, telling someone you don't work on commission is also the best upselling tactic. I often found if I told someone a merino wool tee shirt was worth the extra $50 over the "tech shirt" they were about to buy, they were more likely to hear me out if they knew I wasn't seeing a cut of the sale. Reasonable, considering I had no incentive to lie.

My affinity for wool almost got me into fights with customers a few times. They'd pick up a tech shirt, and I'd reach for a wool one instead, waxing poetic about the function and benefits of nature's miracle fiber.* Only to be told they didn't want to buy any animal-derived products. My inner monologue, which sounded a little De Niro, went something like, "What are you stupid? You would rather buy a PLASTIC shirt that came out of an OIL REFINERY?!" but I had to find a less confrontational way of expressing myself, so I started to keep a printout I made folded in my back pocket. It had a photo of a petrochemical refinery on the top and poufy white sheep grazing on rolling green hills on the bottom. One making wool (nature's perfect fiber) and one making stinky, sweaty, synthetic materials (you might as well wear a garbage bag). I could understand not being able to afford a seventy-five-dollar T-shirt, but I couldn't understand how anyone might think a synthetic fiber was better for the environment than a renewable, natural one—especially one that was superior in function to boot. The photo comparison—I dubbed it "Which Do You Think is Better for the Environment?"—was my nuclear option, and perhaps a dick move. But it always worked.

---

*Wool's illustrious status within the community of outdoor-wear junkies stems from a combination of factors. 1) it keeps you warm when it's wet, 2) it keeps you cool when it's hot out and warm when it's cold out (it has to do that for the sheep who wear it full time), 3) it's naturally antimicrobial so you can wear it for days on end in a variety of conditions and it won't get funky, 4) it's fire retardant, 5) it's natural and renewable. Most "performance wool" comes from merino sheep. The fiber of their wool is finer and therefore light and soft, not scratchy like the thick wool sweaters of yore that most people think of when they hear "wool."

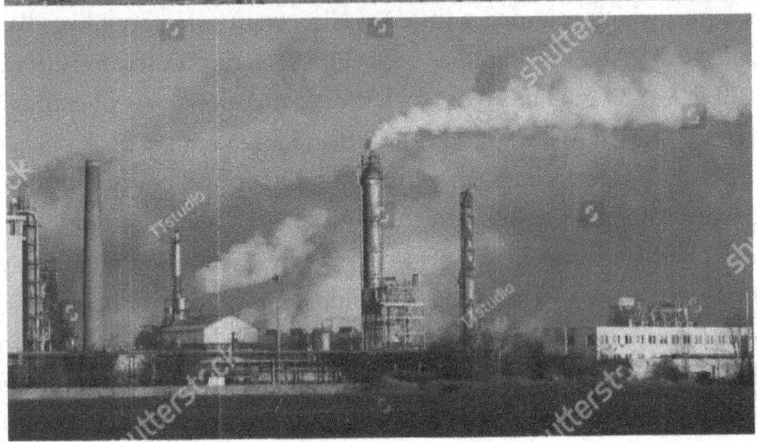

Which Do You Think Is Better for the Environment?

As someone who had worked and participated in the outdoor industry for most of my early adult life, I thought I understood conservation. Ethos like Leave No Trace, organizations like the Sierra Club, taxes... maybe some parking fees? But as I became more involved, I realized how limited this understanding was. One seemingly paradoxical phrase I kept hearing was "hunting is conservation." Or sometimes, "hunting pays for conservation."

It turns out, annual funding for state fish and wildlife agencies–the organizations that pay for research, rangers, biologists, maintenance, etc.–comes mostly from hunters and anglers. Sixty to sometimes eighty percent of their budget is generated through the sale of hunting and fishing licenses and excise taxes on firearms, ammunition, and archery equipment through manufacturers. It's one of the most perfect closed circles of government funding I've ever seen. This insular economic machine is basically a "pay to play" or "if it pays it stays" model. You want to hunt elk? Well, you'll have to pony up, because those suckers are big and need a lot of land to live on, and somebody's got to pay for it. Economists Gretchen C. Daily and Katherine Ellison wrote about the power and necessity of keeping conservation and the outdoors profitable in their book *The New Economy of Nature*. As did former investment banker Mark Tercek and science writer Jonathan S. Adams in *Nature's Fortune*. But you shouldn't have to be an economist or investment banker to understand that Mother Nature has to turn a profit if she wants to contend with gas, oil, mining, and development.

The Federal Aid in Wildlife Restoration Act mandates an 11 percent federal excise tax paid out by the manufacturers of archery equipment, guns, and ammunition. Key Pittman and Absalom Willis Robertson sponsored the bill (giving it its nickname, Pittman-Robertson) and Franklin D. Roosevelt signed it into law in 1937, at a time when American wildlife populations were at an all-time low. The funds generated from the tax get collected by the Department of the Interior, who then distribute the money to the states. There's also a fish-focused version from 1950 called the Federal Aid in Sport Fish Restoration Act that's also referred to more colloquially as Dingell-Johnson after its two champions, congressman John Dingell and senator Edwin Johnson.* What I like about these particular bits of legislation is that they came about because sportsmen demanded the government tax them. You just don't see that every day. I like to imagine it went down something like this:

Hunters and Anglers: Hey we want you to tax us.

Uncle Sam: What? Really? Ok! . . . Why?

H&A: Well, we've noticed a serious drop in wildlife populations all over the country. We think part of that is pollution, crappy farming practices,

---

*My editor absolutely refuses to let me make a Dingell-Johnson joke.

market hunting for meat and hides, plus some other assorted stuff. We figured, if we could pool our money, maybe you could use it to set up programs to buy up land and protect it for the animals, pay for some wildlife cops to prevent poaching, and maybe fund some research into what's happening to the animals, their habitats, and how we can bring their numbers back to what they used to be?

US: So, you want your own private outdoor club?

H&A: No, no. The land should be for everyone to enjoy–we'll just pay for it. We use it the most and recognize we remove some animals from it. So, we want lots of protected land for all the critters to live on. We want *tons* of animals, the way it used to be, Snow White–level, like deer just falling out of trees 'n stuff.

US: Do deer live in...

H&A: No. That was a joke. You need to get outside more.

US: Okay so how do you wanna be taxed? Also, you know about the Dust Bowl, right? And the Great Depression? Nobody has any money. I doubt people are going to be champing at the bit to spend more money on anything right now.

H&A: First of all, do you mean the Dust Bowl you idiots started with your shitty farming practice suggestions? Second, we are the ones asking you to tax us. If we don't do something now, we're not gonna have anything left to save in the future. We were thinking the money would come from state license sales and manufacturer excise taxes on hunting equipment like archery stuff, firearms, and ammunition. You do know what excise taxes are right?

US: Totally. But maybe you should tell me what *you* think they are so I can make sure you're right.

H&A: Well for example, the gun or ammunition manufacturer would pay an 11 percent tax on the goods they make, then they'll factor that into the sale price of their product. This way the customer doesn't even have to see it. It just goes straight from the purchase of that product, a hunting rifle, say, to you to hand out to a Fish and Wildlife department. That way everyone is basically paying to protect and maintain the wild land they're spending time in.

US: And the licenses? We've had those for a while ya know?

H&A: Yeah, but they're kind of all over the place. We want a more uniform system and regulations tailored to the biodiversity of each state. And we still want out-of-state hunters to pay more. If people want to use the natural resources of another state, they should have to pay for that privilege. And we want very clear language that says all funds generated from license sales will go right back to the state's Fish and Wildlife departments—they can't be used for *anything* else.

US: And what's Fish and Game gonna do with all that money?

H&A: We figure every state's department will have unique needs, but, in general, we want them to: 1) buy up, restore, and maintain public wild lands for the benefit of the animals there, 2) make sure the public can access those lands, 3) employ biologists to study wildlife populations and learn how they can be made to thrive again, and 4) fund hunter safety and education programs as well as maintain public shooting and archery ranges so we have places to teach the next generation of hunting outdoor enthusiasts.

US: This all sounds pretty good, and like, obvious.

H&A: Yeah, we know. It took us way too long to get our act together though. No thanks to you, Sam. If we had gotten some more wildlife laws on the books sooner, we might have saved the passenger pigeon from extinction. Better late than never, I guess.

US: I'm gonna put this on the fridge with all my favorite legislation.

H&A: Maybe, one day, other outdoor recreationalists will want to chip in for nature, and we can add a tax to hiking boots, kayaks, mountain bikes, and tents.

US: Let's not get ahead of ourselves.

Okay, so I just crammed over 140 years-worth of hard-fought conservation history into one puerile and reductive conversation between our anthropomorphized government and a group of around nine million people. In truth, the wheels of our conservation funding models, including Pittman-Robertson and Dingell-Johnson, started rolling in the 1830s with the earliest iterations of regulated hunting seasons and the Public Trust Doctrine. In 1850, New Hampshire and Massachusetts established the first game warden programs. In 1875, Arkansas became the first state to outlaw market hunting. Then in 1878, the first bag limit was set, in Iowa for prairie chickens (pinnated grouse). By the early 1880s, every state had at least one law related to game management.

George Bird Grinnell, who (along with Gifford Pinchot and Aldo Leopold) should have as much name recognition as Theodore Roosevelt for his contributions to conservation, was a Brooklyn boy who went to school at John James Audubon's former home in Assigning New York and was taught by his widow, Lucy Bakewell Audubon. She was the early inspiration for Grinnell's life in conservation. Before she died, she bequeathed him a painting Audubon had made of an eagle attacking a lamb that had been Grinnell's favorite growing up. (The painting is now back at Mill Grove in Pennsylvania, a museum of where Audubon first lived in America when he came over from France in 1803.) By 1886, Grinnell had been working as the editor of the outdoor magazine *Forest and Stream*, which would later become *Field & Stream*. That February, he wrote an impassioned editorial on the topic of wild bird feathers being used in women's hats. He titled it "The Audubon Society."

Caption TK

It might sound a little silly, but the millinery trade had become a serious ecological threat that led to the near extinction of a few bird species. Some populations were dropping by as much as seventy percent. The same way wild animals were hunted to be sold in food markets, and bison had been hunted for their robes (pelts), so were birds shot for their feathers to be sold to hat makers. After Grinnell called attention to the issue in *Forest and Stream*, other conservationists who ran in the same circles, like William T. Hornaday (famed taxidermist and bison fanatic) and Frank M. Chapman (ornithologist for the American Museum of Natural History), wrote about the matter for the New York Zoological Society (Grinnell was also on the scientific council with them).

In 1896, Boston socialite Harriet Hemenway read all the grim news about feathered hats, said, "oh shit," and organized a series of high society tea parties with her cousin Minna Hall. They used the parties as a venue to educate other women about the plight of the birds and ask them to help make the hats unfashionable. Harriet and Minna would take up the Audubon Society mantel, which had grown considerably since Grinnell's founding, and start the Massachusetts Audubon Society, which paved the way for the Society to go national by 1905. They gained enough political power to influence state and federal conservation laws including the Lacey Act.

The passage of the Lacey Act in 1900 marked the first federal law to protect wildlife. And though it was meant to offer some respite to passenger pigeons by making it illegal to transport or sell wildlife that was illegally taken, it was difficult to enforce, because it had little funding behind it. Mass slaughter of birds by commercial market hunters was the largest factor in the extinction of the passenger pigeon. The last bird (Martha) died in 1914, but the population, which required huge numbers to sustain itself, was doomed to collapse long before that. Imagine going from fourish billion birds to total extinction in only sixty-five years–pretty fucked up. Market hunting on the whole wouldn't be outlawed until 1918, with the passage of the Migratory Bird Treaty Act.

In 1934, three years before Pittman-Robertson was signed, we got the Migratory Bird Hunting and Conservation Stamp, known today as the Federal Duck Stamp. I'm a huge fan of the Duck Stamp and end all my conservation history talks and museum tours with a plea that everybody buy one. You don't have to be a hunter to buy one–lots of birders buy them too–but you do have to buy one if you plan on being a waterfowl hunter. They cost $25 (at the moment) and for every one of those dollars, $0.98 goes toward the purchase of habitat or a conservation easement for the National Wildlife

As relevant today as it was the day it was published in 1936, Ding Darling's "The Conservation Interests Can Get What They Need If They Will Pull Together" shows how everyone who cares about the environment can achieve greater momentum when they fight for conservation as a team.

Refuge System. Since its inception, the Duck Stamp has been responsible for the protection of over six million acres of habitat for waterfowl, generated over $1.1 billion for wetlands, and created or expanded over three hundred National Wildlife Refuges.

In the 1930's, "Ding" Darling was an avid waterfowl hunter and political cartoonist. While the Dust Bowl dried out the Midwest, Ding used his political cartoons to educate the public about animal and habitat conservation and blast lawmakers for the unsustainable farming practices they promoted and the near nonexistent job they were doing of protecting the environment. His cartoons made such an impact that FDR put him in charge of the US Biological Survey in 1934, which would eventually become the US Fish and Wildlife Service. When the department was kicking off the Duck Stamp program, Ding was tasked with creating the artwork for it. From what he had witnessed during the Dust Bowl, Ding knew it wasn't just hunters who would benefit from healthier landscapes. He used his clout with FDR to organize a conference of conservationists in 1936. The attendees were conservation proponents and outdoor enthusiasts of all stripes, hunters and anglers but also farmers and members of gardening clubs. Ding knew that conservation would benefit anyone who spent time outside professionally or recreationally, so the more diverse a cast of characters he could assemble, the more sway they'd have in Washington. The gathering in DC became the National Wildlife Federation, and they continue to be a trusted voice in wildlife and environmental policymaking. Today, they're one of our most inclusive conservation organizations, with a broad goal of "mobilizing a diverse conservation army" to restore habitat and transform wildlife conservation. They don't focus on one type of habitat or one species, or one type of supporter–they don't care if you're a hunter, a gardener, a kayak polo enthusiast, or a vegan squirrel whisperer. They believe their "goal of increasing America's fish and wildlife populations and enhancing their capacity to thrive in a rapidly changing world" should be appealing to all nature lovers, regardless of what you like to do in the outdoors.

The Wild Duck Chase by Martin J. Smith is a fun trip into the world of "pro-am competitive duck painting" as he calls it. It touches on the culture clash between birders and hunters, but its main focus is on the only art competition run by the US government and selected by a jury–the choice of design for the following year's Duck Stamp. He tells one story, in particular, that I find perfectly illustrates the hunter as obsessive citizen scientist, describing how Duck Stamp artist Robert Bealle "literally knows waterfowl inside and out, having once identified the migration route of a bird he had shot by linking the undigested aquatic vegetation in its crop—an exotic, non-native species called hydrilla—to a particular stretch of the Potomac River. (He also says he could taste the difference in the duck meat when the canvasbacks he hunts began feeding on little clams in the Patuxent River.)"*

Another web-footed phoenix to have taken flight from the ashy detritus of the Dust Bowl, Ducks Unlimited started in the 1930s when founder Joseph Knapp formed the More Game Birds in America Foundation (not exactly a snappy title, but it's no secret what they were about). They published the book *More Waterfowl by Assisting Nature* in 1931, which took a revolutionary approach to migratory bird conservation. The technicalities of what "ecology" was and how best to serve it was still a new science, but the leaders in the field were all following the visionary groundwork laid by Aldo Leopold. One of his teachings emphasized the importance of habitat (generally just referred to as "land") as much as the importance of regulating the hunting of animals. For duck lovers, this was a daunting call to action because waterfowl use migration flyways that can, for example, go from Canada to Florida or Alaska to Mexico. Meaning, they had a lot of land to protect and rehabilitate if they wanted places for their precious duckies to breed, nest, eat, and rest while traveling.

The group was involved in the passage of the Duck Stamp Act, and the following year they organized an international wild duck census that was the first of its kind. Waterfowl population surveys have been the scientific norm for DU ever since. The most significant revelation to come out of the census

---

*The Duck Stamp took a more mainstream center stage in September of 2021, when late night host John Oliver aired a segment on the stamp and its art contest in his HBO show *Last Week Tonight* (season 8, episode 24). It was a fun and accurate portrait of the program (despite Oliver's misidentification of a Brant goose as a duck). His team had five pieces of truly silly duck art commissioned for the prestigious event, but, due to some technical faults and the goofiness of each one, two were disqualified and the other three were voted out of the running. Oliver then put all the originals up for auction and generated just under $100k, all of which was donated to the Federal Duck Stamp Program.

The highest bid for any of the works went to a piece featuring a handsome blue-winged teal flying through a field with the pixilated retriever from the beloved early Nintendo game *Duck Hunt* in the background. *Field & Stream*, whose very roots are tied to the creation of the Stamp, put up a bid of $33,200, with the intention of donating to the worthy

was the importance of Canada's breeding grounds. If they wanted to build up duck populations for the enjoyment of people in America, then they'd have to start by making more ducks in Canada. This meant the organization would have to incorporate in The Great White North if they wanted to buy land there. The fellas had a brainstorm sesh in 1937 about what to call their new guild. Knapp, who now had a track record for such inspiring names as "More Game Birds" and "More Waterfowl," had some equally creative ideas for this new venture including "More Ducks" and just "Ducks." Luckily, his colleague Arthur M. Bartley remembered that the legal designation for organizations incorporated in Canada was "limited" so the group would end up being called "Ducks Limited." Knapp was pissed and shouted "Damnit, we don't want limited ducks!" Bartley responded, "Ducks Unlimited, then."

Ducks Unlimited has become the largest wetlands and waterfowl conservation organization in the world. In 2021, they surpassed fifteen million acres of conserved wetlands. Their campaigns have raised billions of dollars and protected and enhanced millions of acres of coastal and inland wetlands, wetlands that aren't just important to ducks. Healthy watershed systems and wetlands reduce algal blooms and erosion protecting drinking water and preventing floods. And of course, they provide habitat to animals besides ducks as well as outdoor recreation opportunities for people.

Some early wildlife laws and fund-raising measures like the Duck Stamp were seen as merely a way to slow the eventual demise of certain declining species. "Certainly," the consensus was, "the pronghorn antelope was

---

cause and proudly displaying the piece in their corporate offices. I was excited to see this little-known government program receive so much attention from a show like Oliver's, whose audience skews a little younger and more liberal than most of the people who regularly buy Duck Stamps. Oliver noted as much when he chastised the stuffy attitudes of the judges and the short-lived requirement that Duck Stamps must show hunting imagery. It's true that the Stamp was created by and for hunters, but, as Oliver says in the segment, "Potentially driving people away from the Duck Stamp is a really terrible idea. Stamp sales are already down . . . and that means less money for conservation." But what would have generated more money—a winning John Oliver Duck Stamp sold at $25 a pop or the auction that Oliver ultimately held? Either way, shaking up the Stamp competition garnered it more attention, and that was good news for the birds.

going to become extinct." Some people already thought it was, that's how few there were. (Pronghorns are neither antelope nor goats, as their nickname "speed goat" implies. They occupy their own branch on the tree of life–their closest relative is the giraffe. I have a theory that you could show a photo of one around any city outside their habitat range and everyone you ask would guess they live in Africa.) The hope was that by enacting protections, maybe we could stretch out what was left of them to last a few more years, like rationing fresh water on a life raft. In fact, thanks to all the new regulatory practices and the money that started rolling in, pronghorn didn't become extinct. There are more than a million pronghorn antelope living in the United States today.

When Theodore Roosevelt became president in 1901, he ushered in a new national commitment to land and animals that we still laud today. And with the help of our earliest wilderness and wildlife professionals like Grinnell, Leopold, and Gifford Pinchot, the practical aspects of protecting lands as well as species continued to take shape. Conservation is the "wise use" of a natural resource (animals included) so it won't be depleted and can continue to replenish itself in perpetuity. In contrast, preservation is the protection of that resource without any kind of consumptive use. Today there's a slew of public land designations that embrace both methods of protection.

National Parks Preservation. All closed to hunting except a segment of Grand Teton.

National Reserves and Preserves Conservation. Similar to a national park except they're usually open to hunting. Reserves are managed locally and by the state. Preserves are managed federally.

National Forests Conservation. Managed primarily for regenerative timber production. Most allow hunting. Many of them boarder national parks.

Wildlife Management Areas Conservation. While national parks are for "non-consumptive" recreating, and national forests are for timber harvest (among other uses), wildlife management areas are primarily for hunters and anglers. WMAs represent the public land that state wildlife agencies buy and maintain using the funds generated by the sale of hunting and fishing licenses.

Then and now TK

Wildlife Refuges Conservation. Some are state, some are federal. Many are open to hunting, but that's usually decided by the refuge–how big it is, what level of harvest it can sustain, and how much public traffic it gets. If you're confused as to why hunting is allowed in a "refuge" here's how the US Fish and Wildlife Service explains it: "National wildlife refuges exist primarily to safeguard wildlife populations through habitat preservation. The word 'refuge' includes the idea of providing a haven of safety for wildlife, and as such, hunting might seem an inconsistent use of the National Wildlife Refuge System. However, habitat that normally supports healthy wildlife populations produces harvestable surpluses that are a renewable resource. As practiced on refuges, hunting does not pose a threat to the wildlife populations–and in some instances it is necessary for sound wildlife management. For example, deer populations will often grow too large for the refuge habitat to support. If some deer are not harvested, they destroy habitat for themselves and other animals and die from starvation or disease. The harvesting of wildlife on refuges is carefully regulated to ensure equilibrium between population levels and wildlife habitat." What they don't mention is that most Wildlife Refuge lands were purchased with Duck Stamp dollars. If you have ever enjoyed spending time in a Wildlife Refuge, thank someone who bought a Duck Stamp–they bought the land.

National Monument Mix of conservation and preservation. Traditionally this designation was meant to protect specific natural features or those with historical value, like the Grand Canyon, the Gila Cliff Dwellings, or the Muir Woods. But because a president can designate a national monument by executive order, they can protect as much land as they like. If the monument is large enough, they'll allow hunting. Some national monuments, like Zion in Utah, have gone on to achieve the exalted status of National Park.

The National Park Service's Archeology Program has a comprehensive list of all the national monuments with the date they were established and the president who designated them, as well as presidents who have diminished or enlarged them. For example, in 1909, Theodore Roosevelt protected 608,640 acres of Mount Olympus in Washington state. In 1912,

Taft reduced its size by 160 acres. Then Woodrow Wilson reduced it by a whopping 299,370 acres in 1915, and, in 1929, Calvin Coolidge reduced it by a further 640 acres. The seventy-fifth congress stopped the bleeding by redesignating the land Olympic National Park in 1938. Bill Clinton designated the Cascade-Siskiyou national monument in 2000, protecting 52,000 acres. Then Barack Obama enlarged the protected area by 48,000 more acres seventeen years later.

Even though the land might be managed by the Park Service, Forest Service, or Bureau of Land Management, it's the US Fish and Wildlife Service (which was initially the Bureau of Biological Survey) that oversees all the animal science on those lands. And it's the money generated by hunters that gives them the majority of their budget to do it. The Arizona Game and Fish Department tries to clarify that fact in their literature where they have written "There is no alternative funding system in place to replace the potential lost funds for conservation. If hunting ends, funding for wildlife conservation is in peril."

Hunter Education itself is a part of the conservation funding machine. The course costs $15, and 100 percent of that money goes back to state wildlife agencies. The people who teach the classes and administer the exam are all volunteers; I'd like to become an instructor myself sometime. But in terms of hunting and how it pays for conservation, Hunter's Ed is the tip of the iceberg, because it's graduates of the program who make the biggest contributions to a state's conservation efforts. My first hunting guide, Fisher Neil, runs a small-scale, urban-based operation, taking people out for whitetail deer and small game. In 2019, the money generated just from his client's license sales came to $6,800. In 2020, that number rose to $10,006. (Hunting license sales jumped all over the country in 2020–2021 due to the Covid-19 pandemic and people suddenly having the time to get back into the woods or take to the field for the first time.) All that money went back to the New Jersey Division of Fish and Wildlife. Adam Gall, an elk hunting guide in Colorado who runs Timber to Table Guide Service with his wife, Ana, told me their clients generated $15,000 in conservation funding just through license sales in 2020. That all goes back to Colorado Parks and Wildlife.

If I, like so many other hunters, hope to forage for meat in the wild–the mountains of Colorado or the suburbs of New Jersey or wherever–then

I've got to pay to keep those areas a suitable habitat for all the animals that live there. And if there isn't as much habitat, then I'm basically paying the wildlife agencies to let me manage populations on their behalf. Otherwise, the agencies would have to perform the same service themselves to keep populations healthy—and they'd charge the taxpayers to do it. If I buy some grocery store burger meat at the market instead of a deer tag, my money will go to a bunch of different places, but not one of them would be the woods. As much as I love my farmers markets and the small-scale meat providers I got to know through the butcher shop, there's just no arguing with the superiority of a system that generates meat, jobs, and funding for the environment by keeping nature wild, healthy, and valuable.

As I learned more, the idea that "hunting is conservation" started not only to make sense but also revealed another side to the phrase. The proud statement that hunters pay for conservation is a subtle reminder that everyone else using (and profiting from) the land and water does not. A "backpacker's tax" has long been debated and occasionally attempted. It would function in the same way hunters and anglers contribute to conservation through the 11 percent Pittman-Robertson excise taxes on related hunting and fishing products but would be added to outdoor and boating equipment like technical backpacks, trekking poles, mountain bikes, climbing gear, kayaks, you name it. Hiking boots can cost anywhere from $90 to $400, but let's say the average pair costs $150. Eleven percent of that would be $16.50. You could pledge to donate that amount to a conservation organization every time you buy a new pair of boots. But will you actually do it? I doubt I'd even think to. When the money is built right in, it's more effective. Set it and forget it.

REI loves to talk about how they give back to the outdoor community. When I worked there, I was always proud when I heard the latest numbers roll in about their corporate giving. In 2020, REI invested $6.3 million in its 400-plus nonprofit partners. But the company generated $2.75 billion that year. Just half of 1 percent of that would come out to $13.75 million. That 6 mil' doesn't sound so impressive anymore, does it? Yvon Chouinard, the founder of Patagonia (which REI sells), started 1% for the Planet which, at its most basic level, asks member companies to donate

1 percent of gross sales directly to environmental nonprofits. "This is not philanthropy," he says. "This should be a cost of doing business. It's paying rent for our use of the planet." An organization called 2% for Conservation upped the ante by asking its members to donate 1 percent of their money (which is structured differently if you choose to be an individual or corporate member) and 1 percent of their time by volunteering in the conservation space. Any company can join, but the majority of members are from the hunting and fishing community.

	Kayaking in New York State (Resident)	**Hunting in New York State (Resident)**
safety training or license requirements	None	Hunter Safety Course $15 General Hunting License $22
necessary gear*	Kayak $700 Paddle $150 PFD $120 (please don't buy cheap safety gear)	Rifle $900 ($99 is excise tax) 20-Round Box of Ammunition $41 ($4.51 is excise tax)
total spent =	$970	$978
money that went to conservation =	$0	$140.51**
recurring costs that support conservation =	$0	$22 for renewing your license, plus the excise tax on all additional ammunition**

*outdoor gear can be a little less or a lot more expensive than this

**plus whatever $25 or so you might have spent on a Duck Stamp or other habitat stamp

Can you imagine how much money could be generated for the environment if every outdoor goods manufacturer was held to the same financial standard as hunting and fishing gear manufacturers are? But outdoor gear and apparel companies keep saying no to the tax. They say, "outdoor gear is expensive enough." And that's why hunters and anglers get their undies in a bunch over "hunting pays for conservation." It gets old pretty fast to hear flowery speech about folk's love of the outdoors while, in the same breath, they decry the hunters and anglers footing much of the bill to keep wildlands protected for everyone.

Outdoor recreation can complicate the wildlife preservation equation. You might reason that "non-consumptive" recreation activities don't impact wildlife as much as hunting–activities such as hiking and mountain biking might seem less impactful to animals than hunting them, but when you look more closely, it's not so simple. Both require trails to be built through animal habitat and can change animal behavior. Colorado is good example of this dynamic. As remote work became more commonplace in the 2010s, white-collar workers took the opportunity to relocate closer to their preferred recreation spots, and Colorado was one of the western states that flooded with new full-time residents. The Covid-19 pandemic exacerbated the trend. Outdoorsy folks could now hit the slopes in winter (when there's enough snow) and hike or mountain bike all summer (when the forests aren't ablaze).

Mule deer, elk, and whitetail used to escape development to find safety in the mountains, but they now find pressure from increased human activity and development in more places and in all directions at all times of year. I asked hunting and fishing guide Adam Gall his thoughts on the issue, and he told me, "The [mule deer population] peak that we saw in the early 2000's and the objective that biologists hope to regain statewide [in CO] may not be a realistic number anymore. There's just too much pressure from human development resulting in habitat loss."

Cities continue to spread out into what was habitat. Denver is particularly guilty, as its "view plane rules" restrict vertical development in order to preserve citizen's views of the mountains. Because the rapidly growing city isn't allowed to build up, it builds out instead. This means more sprawl, more driving, and less habitat. The growing pains echo those found in other western states.

In a 2016 issue of *National Geographic* that was cover-to-cover about Yellowstone and the Greater Yellowstone Ecosystem, Todd Wilkinson wrote, "The economy and culture of the region are changing fast.... The Old West economies, defined by mining, forestry, and ranching, are stagnating, while Bozeman and Jackson are growing like weeds." For the lifestyle pilgrims moving to those boomtowns, remoteness is no longer an obstacle; they can telecommute in the morning and fly-fish, hunt, or ski in the afternoon. "People want to live and recreate next to spectacular, largely unspoiled landscapes," says economist Ray Rasker of Headwaters Economics. "The thing we need to figure out is how to deal with an unprecedented wave of newcomers and not turn Greater Yellowstone into the places they fled from." Everyone wanting to look at nature can make it harder to have nature around to look at.

Mountain biking is becoming an inflection point in Colorado. It has no excise taxes on related equipment meant to offset its impacts on the environment, and the industry built around the sport benefits when more trails are developed. In an article for the mountain biking site *Singletracks*, author John Fisch explores the recent trend of private landowners developing mountain biking trails and describes their specific appeal by stating, "landowners have greater freedom to develop trails on their property, free from public hearings or environmental reviews. As a result, trails are built specifically with mountain biking in mind, and can incorporate features that would never be approved for public land." In other words–no oversight and no considerations for habitat.

In contrast, the hunter prefers there be no alteration to the landscape and pays to re-wild or expand habitats, not develop them. Hunting license dollars pay for biological studies, and the industry built around hunting only benefits if habitats expand and species numbers are strong. The hunter might get lucky and shoot an elk, removing it from the landscape, but elk reproduce. Habitat turned human playground does not.* None of this is meant to pit cyclists against hunters, but it should serve as a reminder that there is no such thing as a non-consumptive outdoor activity. Instead of being categorized as consumptive or non-consumptive, activities should be rated by their impact on habitats and how that informs various species' ability to thrive when they have to share the landscape with that activity.

Outdoor recreation is big business, and, to the extent you need the outdoors to have an economy based on it, that's a good thing. You can't buy sustainability in a store, but you can pay into activities that require it. Headwaters Economics calculated that "In 2020, outdoor recreation contributed $374 billion (1.8%) to the nation's gross domestic product. This is more than twice the size of motor vehicle manufacturing and air transportation, three times the size of oil and gas development, and nearly four times the size of performing arts." The revenue brought to the Department of the Interior just from hunters and anglers came to $1.5 billion in 2021. So how nitpicky should we be about one outdoor activity versus another? If we set aside enough land for the skiers, mountain bikers, and animals, I don't think it would matter. It's the having enough land to do

---

*Colorado's situation is complicated further by a new initiative to reintroduce wolves to the state. Humans removed wolves from Colorado and the rest of the Continental United States in a sadly impressive feet of localized extinction in the late 1800s, but the resilient wolves have rebounded since then, coming back to the United States from Canada and Mexico on their own and through reintroduction programs. The absence of wolves took a real load off elk and deer (and lead to dangerous overpopulation in some regions) back in the day; their reintroduction to the landscape will add an additional pressure to today's elk and deer populations already burdened by human expansion.

Wolves bring with them a trophic cascade that comes from the landscape of fear they generate. The landscape of fear is how wolves scare the pants off deer and elk and how that fear affects their daily movements. The trophic cascade (an avalanche that starts with apex predators) is how those movements end up affecting habitats and other animals on a broader interconnected scale. The cascade is fascinating but often given total credit for environmental changes that should be attributed to a variety of factors occurring in concert. It's tough to calculate the rate of future fawn mortality from wolf predation and the toll that will take

that part, that worries me. And right now, there's no system in place to ensure that outdoor recreation industries that profit from the land also give back to it.

---

on Colorado ungulate numbers (in general, western wolves prefer elk to most everything else, but all calves and fawns are on the menu in spring). Maybe the wolves will have a meeting and decide to cut struggling mule deer populations some slack by only eating whitetail. (Wolves also come with baggage and controversy among the rural communities who would be sharing more space with them than the urbanites who most often vote for their reintroduction; one of many familiar schisms making up the lamentable rural-urban divide.)

Some Colorado ranchers make extra income by letting people hunt on their land for a fee, which incentivizes the ranchers to keep land as habitat. But if deer and elk populations drop due to wolf reintroduction and pressure from outdoor recreation-related habitat loss, ranchers might not be able to make money that way anymore. Instead, they could just as easily sell off parcels to developers eager to cash in on ever-expanding human populations.

3

The Taxidermist

History, Heritage, and How to Talk about Death

He jumped and kicked his legs back in the air then went crashing through the brush, and we heard an abrupt quiet. It's very rare that you shoot a big game animal and they just drop stone-dead to the ground. Adrenaline can push them a hundred yards, even with a perfect shot to a vital area of their body, ideally the heart and lungs. Though I wished it wasn't the case, the moment took longer in my head than in reality. "Perfect!" my guide said, certain it was a heart shot, which meant we could climb down from our tree and start tracking him right away. I just sat there quietly for a moment and said, "I didn't think it was going to happen."

When you're bow hunting, you have to get a lot closer to your target to make a good shot than if you're using a rifle, so you should have a good idea of where your shot lands on an animal right away, but you can confirm further by finding your arrow (or bolt for a crossbow) on the ground and looking at the color and consistency of the blood on it. Based on the color, you can tell where the arrow passed through (and it does pass through entirely if it's a good shot—it shouldn't stay sticking out of the animal). Subsequently, that color will tell you how long until you should begin tracking. Bright red and bubbly means heart and lungs, and you can start tracking right away; viscous dark maroon means liver, and you should wait around three hours; yellowish or greenish and flecked means a gut shot, and you should wait up to six hours to begin tracking.

I remember seeing hunting shows as a kid and always being angry when the hunter would take their shot then say something like, "We'll give him a few minutes then head down." My assumption was that they didn't want to take the chance any of the animal's last mournful death throes would be captured on film, and I resented their cowardice at wanting to be a part of the killing but ignoring the dying. Yet as I learned more about tracking, I found the truth to be both more logical and humane (a theme that reoccurred while I learned more about hunting). All the aforementioned shots will be fatal, but by giving the animal space to lie down and die on its own time, you save them the stress of a predator following them. If you follow them too soon, you might push your quarry farther and farther away from you, scaring them unnecessarily and running the risk of losing them, and the meat you hoped to harvest, entirely.

There's no secret ritual or blood rite needed to become a hunter. Take Hunter's Ed, get a license and ask for help. You don't need a background in the outdoors or a shelf full of Hemmingway, Capstick, or Ruark; you don't need the latest gear; and you don't really even need to know what you're doing. It's okay to be a beginner. I love being a beginner, because it means I'm learning something new. That aspect of hunting became an instant attraction to me. No matter how long you've been hunting, it's always different. Because even if you hunt the same woods your whole life, the animals you pursue are responding to you as much as you are to them. I found this refreshingly different from when I was a white-water guide–then, I responded to the river, but the river never responded to me.

I made mistakes when I started hunting by myself, and I have no doubt I'll continue to make them into the future. Having learned to shoot sporting clays from a young age, I came to hunting as a pretty good shot and with an abundance of confidence in that arena. I figured that if I could hit a small, fast-moving target, then an even larger stationary one shouldn't be that difficult. Logically, it's not. But I hadn't considered the heart-pounding emotional aspect of aiming at a target that was alive, and how that would color the shots I chose to take or pass on. I've passed on shots I probably could have taken, but I'd rather make that "mistake" than to take a shot I'd wish I hadn't. Once, I shot a deer whose antlers, upon closer inspection, didn't meet the restriction requirements of the unit I was in. I could have lied on my tag report (which, to be clear, would be a crime), but I reported him to the local game warden. I was embarrassed, but, thankfully, I was allowed to keep him. As a new hunter (and still to this day), I read a lot about hunting and watched endless instructional videos, but the best lessons I received came from guides on hunts or hunting with friends. It's not that the information is so different–there's just no substitute for being out there. When I wasn't outside practicing, scouting, or hunting, I'd find myself at my computer with nine different tabs open, deep diving into whatever topic came next on my list, trying as best I could to prevent the next mistake before it happened.

While I benefitted from inherited wisdom from myriad sources in books, person, and online, I also found benefits to being a beginner–coming to an experience with no preconceived notions means you get to look at everything with fresh eyes. Different terrains call for different hunting styles, and I knew that, in the heavily forested Northeast, I'd probably be hunting from a tree a lot, so I took my time researching all the different tree stand options. A climbing stand piqued my interest initially–I liked the technique of pulling the bottom of the stand up the tree with my feet before hoisting the top portion and repeating, like a camo-clad inchworm working my way to a desired height. But the climbing stand required too many variables of the tree, limiting its versatility, so I scrapped that idea. A ladder stand was also out. They're too big and take too long to set up–they're "portable," but I wouldn't call them mobile–not a good choice for anyone interested in hunting public lands. A hang-on stand is perhaps the most common type of portable tree stand (it's what I would use in some of my first guided hunts), but the bulky equipment and extra weight was unappealing to me. The moment I saw tree saddles in action, I was smitten. They're basically an arborists tree-climbing harness. Where hang-on stands require you to schlep around a platform to sit and stand on, as well as a safety harness so you don't fall from the tree, saddles act as both seat and safety harness, requiring only a small platform the size of a laptop to put your feet on or against in any way you feel comfortable. The versatility of positions I could safely and comfortably take from a saddle would broaden my field of vision and shot opportunities. It seemed like a no brainer, but when I started researching brands, I realized tree saddles were a hot topic. My searches turned up dozens of articles and videos that essentially broke down to a younger hunter extoling the virtues of tree saddles to assuage the skepticisms of an older crowd.

My beginner status did not make me immune to picking up habits I had always rolled my eyes at in old hunters. While on vacation in Chicago one February, my vegetarian, very-much-non-hunter friend made fun of me for wearing a camo jacket around. My wardrobe has always been dominated by wooly forest greens and browns, but I had found myself wearing camo on occasion of late. The jacket was designed for hunters and was lightweight, packable, windproof, and extremely warm, which also made it the perfect jacket for a traveler. It had a hood too, a feature I don't like on my "around town" clothes but I very much wanted to protect my neck in the Windy City (they'll tell you it got the name because of the city's political history, but it's a double-entendre for a reason). While bringing the jacket was a practical choice on my part, I began to see a glimmer of the culture I hadn't understood before. It could be that you just spent a bunch of money on a new technical jacket, and you want to get some off-season use out of it, but it was also a symbol of community for a proud minority. The same way my fellow Cubs fans wore their blue-and-red hats and jerseys, I enjoyed striking up conversations with the people who started to ask me "are you a hunter?"

Hunting is the oldest tradition or heritage anyone on this earth can claim. Evolutionarily speaking, humans would not be what we are today without meat eating and hunting. It allowed us to walk upright and develop our intelligence. You can't win a fist fight against a wooly mammoth, and nobody decorates a pot or paints a picture until they've fed themselves first. The relics we have of our earliest ancestors are the remnants of their hunts, the stone arrowheads and spearpoints they used while hunting and the cave paintings they made to memorialize the hunt after it was over. I don't think the words *heritage* or *tradition* go deep enough linguistically. We literally evolved our brains and bodies to hunt better—we did not evolve to wear specific religious accessories (heritage) or shoot off fireworks on the Fourth of July (tradition). Whatever particular heritage or tradition you hold dear, hunting is deeper, older, and a part of all of us.

Humans are pretty lame compared to the rest of the animal kingdom. We don't have prehensile tails, we aren't that strong, we don't have claws—we can't even shoot blood out of our eyes. Because we were so uncool, we had to develop at least a few features to help us survive. The trifecta of our superpowers is walking upright, big brains, and sweat. Our eyes face forward like other predators, but because we couldn't be guided by our noses (weak sense of smell) or our ears (mediocre hearing at best) we needed to put our eyes up higher so we could see things farther away—walking upright solved that problem. Getting nutrition from vegetation alone takes a lot of bodily effort—meat provided more calories for less digestive work, and we spent those energy savings developing our brains so we could devise easier ways for our slow, weak bodies to acquire dinner. First, that looked like throwing rocks. Then pointy rocks when we figured out spears, which also made those rocks fly farther, and finally bows, which could throw the rocks for us. Sweating is my archnemesis because I have a nervous system disorder that keeps the faucet running longer than most people's, but for those humans unencumbered by hyperhidrosis, the ability to cool ourselves while expending energy meant we could apply slow-and-steady methodology to our hunting tactics and become persistence hunters. We couldn't fight our prey, and we couldn't outrun them, but we could outlast them.

While all this is not an inaccurate way to frame hunting, it's not a great defense of it. Dubbing something *heritage* or *tradition* is often a thoughtless way to endow a free pass—and hunting has more to offer than that. I want people to understand why it's something worth doing today and worth keeping around even if you don't participate in it yourself. Its value is greater than the fact that it's old. Some people treat heritage as a thing you're not allowed to question because it's part of someone's personal or cultural history or tradition. Bullfighting is claimed as a historical, cultural heritage by its proponents, but once you start to examine it, you see that's really all it offers. The prolonged prodding of an animal for the pleasure of paying spectators would be a tough sell if you were trying to invent

A Very Brief Timeline of People Hunting

2 million years ago	Evidence of the earliest hunters, *Homo erectus*, the direct ancestor of modern humans
500,000 years ago	Earliest use of stone-tipped spears for hunting
315,000 years ago	Earliest representatives of our species *Homo sapiens*
45,000 years ago	A cave painting of a boar in Indonesia is the earliest known painted image of an animal
12,000 years ago	The beginnings of agriculture and animal domestication
9,000 years ago	Organized big game hunting by men and women
5,500 years ago	Humans domesticate horses (this is a big deal)
4,600 years ago	Earliest example of wildlife management on record: hunters and fishermen were taxed one tenth of their take in ancient Egypt
260 years ago	Industrial Revolution marks the start of a growing human separation from nature

bullfighting today. The value of any group's heritage or traditions are often only valuable to those who practice them, and they may cling to the untouchable nature those words bring, knowing that alternatives sound cheap. Just because your great granddaddy's granddaddy did something, doesn't automatically mean it's okay for you to do too. If you want something to stick around, it must maintain or increase its value.

The words *sport* and *hobby* are also imperfect descriptions of hunting. Some people feed themselves and their family exclusively on wild game meat. Some became hunters to connect more deeply with the outdoors and become a part of the wild food chain rather than contribute to a manufactured exploitative one. And yes, many take immense pride in their personal hunting heritage—not the fact that Cronk was a hunter a million years ago (although that is pretty cool), but that their parent was a hunter and brought them into an outdoor lifestyle that they can now share with their own children (something that's becoming more important to parents who see their kids getting farther from nature each day). Hunting doesn't belong in the same category as tennis. The confusion comes from the term *sport*, which has a different meaning in relation to hunting than it does to games in which you whack a ball around. It was the name chosen to separate what we know as hunting today from the market hunting practices responsible for declining species populations of the past. It's not *sport* as in a game; it's sport as in *sporting*, being evenly matched with your quarry and giving the animal a fair chance at evading you, the same way they would any other predator. This is something animals were denied when market hunting practices and tactics were the norm. *Lifestyle* is the closest and best replacement for *sport* and *hobby*, but I think it's still inadequate. Ask any hunter, and I bet hunting occupies a significant part of their life, either in time or memory, and, for a lucky population, both.*

---

*At some point in our friendship, Kris De la Torre told me a story about hunting, communication, and tradition that really stuck with me. When she was a kid, some family members returned from a hunt with a deer in the bed of their truck and chuckled while they told her it was Bambi. The taunting really did a number on her, and it turned her away from her family's hunting heritage all together. When I asked her how her feelings about the incident have shifted over the years and how she thinks about it now, she gave this thoughtful answer.

"I grew up around hunters and amongst people who tell stories about eating squirrels and raccoons in West Virginia as a point of pride. My feelings about it are more complicated now than they were when I was a kid. When my stepdad and his friends taunted me with "Bambi" in the back of their truck, I was freaked out because of the tenderness I felt for animals generally and for the personified, long-lashed, Disney-fied orphan that was Bambi specifically. I definitely think it was over their heads that that movie was on heavy rotation in our living room, and that it positioned the hunters (a.k.a. people who were my protectors, role models, and caregivers) as the bad guys. I know they were making what they intended as a harmless joke and did not mean for it to land the way it did.

I met my friend John when I first moved to New York and we were both working for Eastern Mountain Sports. Between our shared interests in the outdoors, history, design, and New York–style misanthropy, we became fast friends. He was easily responsible for 90 percent of the times I got into the woods after moving to the city. We would go on multi-day canoe trips in the Adirondacks or shorter overnights to Bear Mountain or Harriman State Park, and to camp with him is to observe a master who has perfected his craft. On one such camp-out, I woke up to the smell of fresh bread being baked over the fire, and by the time I stuck my head out of my tent, he had already used a beaver-tooth-shaped hook knife to carve a bowl to put the butter in. I requested this service on all subsequent ventures.

"In years since, I think what was truly sad in that specific moment was the disrespect and attitude of dominance they translated to me about our relationship to the natural world. It's particularly unfortunate because, in recent years, I've had conversations of increased nuance with my uncle and grandpa about what they enjoy about hunting, and what is in their hearts is actually very, very different than arrogance or cruel humor. The words they use and the words I use are different, but I've come to appreciate that they love that family time together. They love setting up camp and quietly moving through the woods. They love the self-sufficiency of filling their own freezers with something they hunted and sharing the abundance with their neighbors and friends. Even those stories of squirrel hunting in West Virginia, which embarrassed me when I was young, now feel like an important part of a family and regional culture with which I don't have any other connection. The wildness of nature where those people lived is something that I don't think I'll ever get a chance to experience.

"The last few years my pawpaw, uncle, and cousin did extend offers to bring me along on a hunting trip. Even if only to camp and sit around the fire. I never made it out with them. Three years ago, my cousin took his own life, and last year my pawpaw passed away from a surprise diagnosis of stage four pancreatic cancer. I can only imagine how sacred and precious the memories of those hunting trips must feel for my uncle now. I don't like to think about how it might hurt him if I asked to hear those stories without his fellow narrators to elaborate alongside him. Before we lost my cousin and pawpaw, the three of them would tell these hunting stories in a sort of waterfall style. One would elaborate on a certain part or ask for clarification or finish the sentence of the other. They would tease each other about the parts of the stories that were accidentally or intentionally omitted, until you came away with a full story and a subtle sense that they're also sharing some inside joke, not to be revealed to the non-hunter. On one hunting trip, the three of them stuck it out in blizzard conditions in a hunting cabin with very few supplies. Their spirits stayed high despite the snow piling up in front of the doorway and the fact they didn't end up shooting anything that season. I love that story and I hope my uncle feels compelled to tell it again someday."

I could tell John loved camping more than I did, and it made me crazy. I couldn't understand why my enthusiasm didn't match his. Eventually I figured it out, and for a gear-slinging outdoor guide, it's a bit ironic–I hate hiking. As much as I love the outdoors, and as much as I love to spend hours walking through a city exploring, I have never liked hiking. I never understood the point. You just walk up a path, then turn around and go back down it? How do you get better at that? How do you make that more interesting? When John and I would venture out, hiking was just our transportation to a campsite, not our end goal. And while the activity of setting up camp and prepping bear bags and firewood was good for keeping my hands busy, and I wanted to improve my bushcraft skills to John's level, my mind kept wandering. Why was I out there? I loved wild animals and wanted to be where they lived, but I hardly saw any when I'd go hiking or camping. Most skittish wild creatures aren't enthused about noisy humans rattling around in their house and tend to steer clear of human-utilized hiking paths. (Though there was that one night we accidently locked a field mouse in our food bag–best night of his life.) Camping with John offered me more than just hiking and setting up camp. I'm sure he orchestrated our trips that way for my benefit. Like an overly energetic dog, I always need an activity, or I will rip the metaphorical couch pillows to shreds. I've only ever been able to tolerate going to the beach if I had something to do there like snorkeling or surfing (or my dream of combing the sand in a Hawaiian shirt with a trusty metal detector and wide brimmed straw hat). John could relax in the woods the way my sister could relax on a beach. But I needed something more.

As soon as I took up hunting, I had an awakening that felt as much like relief as it did anything else. This was different from the relief I would have after my first "successful" hunt (I'd go hunting seven times before I finally connected with an animal)–it's something I felt on my very first hunting trip, before ever even seeing an animal. I had found new purpose in the woods. A reason to be up that early, a reason that activated me beyond my love of looking at trees. I'd feel that joy of purpose every time I went out, regardless of what I came home with. I had found a way to be a part of nature, not just a spectator, and I realized that was something I had been missing and craving. The camping trips I'd always found most fulfilling were the ones where I'd seen an animal or brought home seasonal wild treats like fresh greens or mushrooms. Hunting was just a focus on these two goals: seeing animals and bringing home seasonal foods.

Every time I hunt, I delight in how it requires me to engage all my senses. There are so many things I have to do to be successful, and even if I do everything right, that still doesn't guarantee success. I have to move quietly and with purpose. I have to look in the right places for the right clues. I have to know what the clues mean. Physically, I have to put myself in those right places, and mentally, I have to figure out what those places are and strategize how best to get to them. As a hunter, you have to truly understand the environment you're in, and the ecology around you; you can't just traipse through it. I finally had a use for all that stuff I'd been learning about animals since I was a kid—instead of being an armchair biologist's party trick, it was now practical knowledge.

The year suddenly became a giant advent calendar. I was reminded of the edible bounty that came with the seasons. It's easy to forget that food is seasonal when the local grocery store carries fresh fruits and vegetables from all over the world year-round. I now had seasons of wild game to look forward to, and there was no buying that. Turkey in spring, trout in summer, deer in fall, and small game in winter. If I wanted to stock my freezer with these seasonal delicacies, I had to (and wanted to) pay more attention to the woods and what stories they were telling throughout the year. I had to earn those dinners. I had to get off the trail. The subtle clues left by wildlife in the forest started to look like neon signs, and every walk in the woods became more exciting.

This phenomenon is far from unique to me. In his book *The Mindful Carnivore*, Tovar Cerulli documents how taking a more active role in the surrounding environment changed his perspective on it. "Now that I was fishing, the water had come alive. Ponds and lakes were no longer mere scenery. When I walked along a brook or drove over a bridge that spanned a river, I wondered what fish lived there. . . . Water was no longer just a surface to glance at or paddle across, but a living depth to participate in." Lily Raff McCaulou says much the same in *Call of the Mild*, "When I drive over a bridge, I glance at the water... Now a river catches my eye because I know how to decipher some of its secrets. Fly fishing is teaching me how to imagine some of the life that takes place beneath the water's surface."

As an animal nut, I had always wondered what fish swam below the surface of the water, but that was where my curiosity ended. The more I hunted, the more the entire ecology of a place and its wildlife became special and exciting to me. The connectivity of it all was fascinating but also a tool. The more I could understand the ecology of the area the better hunter I'd become. That type of investment in education, time, experience, and interaction bred a connection and custodial passion I'd never felt in any other form of outdoor experience. You simply cannot look that closely at nature's web without getting caught up in it yourself. Roosevelt's words from *Outdoor Pastimes of an American Hunter*, over 120 years old now, might connect my feelings to those of Cerulli's and McCaulou's best: "It is an incalculable added pleasure to any one's sense of happiness if he or she grows to know, even slightly and imperfectly, how to read and enjoy the wonder-book of nature. All hunters should be nature-lovers."

If animals have been my number one interest in life, history has always run a close second. I didn't have a strong cultural or familial tradition or heritage to speak of, but my grandmother owned an antique shop, and my love of history grew from the stories she would tell me about the objects in her store or at estate sales. I'd bring something over to her, and she would tell me what it was, how it was used, who made it, and what eventually replaced it or made it obsolete. Once, I found two small silver figures in her shop, the head of a bull and a bear. Bears were always my favorite animals, and I loved bulls because my favorite childhood picture book was *The Story of Ferdinand*–I thought it was mere kismet to find these two noble creatures represented in miniature statuary. After begging my grandmother not to sell them, I was given a lesson on Wall Street's dueling mascots and the necessity of paperweights in a pre-air-conditioning open-window world. Antique stores were like museums where they let you bring home the stuff.

I think it was the interplay of fantasy and reality that attracted me to history– (history buffs are susceptible to affection for times and places they'll never get to see) the romance of all the cherry-picked good stuff and the "true story" horrors that I could gawk at from the safety of the present, like learning about how Civil War amputations were carried out. As I got older, I looked for jobs where I could incorporate both my love of animals and history. In high school, I juggled two such jobs in Charleston. One as a groom for a historical carriage tour company and the other at the South Carolina Aquarium, where I gave the daily Alligator Talk. While holding cute, squishy peeping baby alligators and talking about gators and their habitat and the differences between alligators and crocodiles, I liked to wedge in the legend of President John Quincy Adams' pet alligator, as well as a few stories about Charles Waterton's wrestling of a caiman crocodile in Guyana.

I had become infatuated with Waterton, an eccentric British naturalist and taxidermist, after seeing the most excellent painting of him at the National Portrait Gallery in London with my dad. The word *eccentric* was basically invented to describe this dude. My favorite Waterton Fast Facts are: 1) he taught taxidermy to John Edmonstone, a freed slave who would become Charles Darwin's taxidermy instructor; 2) his dog was the foundation for the modern English mastiff; 3) he turned his own property into what would become the world's first nature preserve; 4) he fought a soap factory in court for polluting the park and lake that surrounded it, the first legal battle of its kind over industrial pollution; and 5) instead of drawing political cartoons, he created political taxidermy art pieces. He did other normal things too, like dressing up as a scarecrow and hiding in trees and pretending to be a dog and biting guests when they came into his house. I don't use the word "hero" lightly but…

The portrait in London dates from 1824 and depicts Waterton looking off into the distance with the faintest hint of a smile that's both regal and a touch smug. A red cotinga from Guyana is perched on his finger and looks up at him, with what I see as adoration (it was painted from a mounted

Portrait of Waterton, painted by Charles Willson Peale (1824)

specimen that is still in pristine condition today thanks to his skillful work as a taxidermist). There's also a mounted cat head sitting on a book and looking right at you, no matter where you stand. It was painted by Charles Willson Peale, who was a respected portrait artist as well as a naturalist (among even more talents–a true polymath). Peale established the first museum in the United States, in 1786 in Philadelphia, and used Linnaean taxonomy to categorize the collection, rather than just tossing stuff on the walls where it might look cool.

Peale's self-portrait, "The Artist in His Museum" (1822), is another favorite painting of mine (and of others who drool over museums and natural history), though I only made the connection to Waterton years after seeing his portrait in London. Peale's self-portrait shows him standing at one end of an impressive gallery in his museum known as The Long Room, holding up a red curtain with one hand while the other is outstretched, palm up, beckoning the viewer to come inside and have a look. The gallery patrons are, most prominently, a woman in a yellow dress, a little boy with his father, and a single man further to the back. Peale is quite literally lifting the veil from science, natural history, and art. He asks us (no matter who we are) to join him in the quest for knowledge and insists that an education in science and nature benefits society as a whole and is meant for everyone, including women and children, a rather progressive philosophy for the time. Other notable goodies in the painting include walls of mounted birds in the background, a partially or soon-to-be mounted turkey resting on a box of tools, and the museum's most prized possession, the skeleton of a mastodon. This was the first complete skeleton of a mastodon discovered and assembled in America. It was found on a farm in New York in 1801. Peale joined the expedition to unearth the rest of the skeleton and painted a rad depiction of the event afterward titled "Exhumation of the Mastodon" (1806).*

---

*I love natural history museums—they're the first thing I search for when visiting someplace new. So, I'm extremely lucky to spend as much time as I do at the American Museum of Natural History in New York City. I give a four-hour walking tour of the habitat dioramas there (and I consider that rushed.) An exchange I hear on occasion while walking through the halls goes something like this:

"Wow they're so beautiful!"

"Yeah, but they're all dead."

I can only roll my eyes so many times before they become stuck, so here's why I'll take museum dioramas over zoo animals in a fight. Let's skip over the well-worn, heated debate over the pros and cons of zoos in general and get straight to education. You won't see animals exhibiting a true range of natural behaviors and interacting with their natural habitats in a zoo, because they aren't in a natural environment, they're in a zoo. Do lions spend hours sleeping and licking their balls in the wild? Yes. But that's not what inspires awe. Dioramas can play the hits and take you to places zoos can't. Diorama artists can put you high up in the jungle canopy or on the expansive plains of the savannah, and they can show you, up close, exactly how lions hunt or mandrills teach their young to search under rocks for grubs (cont.)

I started collecting taxidermy when I got to college. I wanted to live in my own natural history museum. It never bothered me that the majority of my collection came from hunted animals. I figured I wasn't the one who had killed them, and I had a sort of Island of Lost Toys feeling about them. For one reason or another they were no longer on that hunter's wall or in that museum's collection, and I felt bad for them. Someone had gone through the time and expense of having an animal mounted and now it was homeless and for sale. I was especially drawn to ratty old pieces that were either poorly mounted (i.e., ugly) or were deteriorating because they hadn't been cared for properly. I worried they might not find homes if I didn't take them in. I'm less prone to anthropomorphizing things these days but only slightly less sentimental.

Five or so years of collecting and obsessing later, I met Joanna Ebenstein. My friend Kris had pulled a magazine article for me about her personal collection of books and oddities dubbed the Morbid Anatomy Library (after her blog of the same name) and I was immediately delighted and envious of this kindred spirit who had managed to turn her own collection of rare and unusual books and artifacts into a mini public

---

on the forest floor. It's a life-size 3-D still from a nature show, happening right in front of you. Those are the behaviors that inspire awe in kids, and it's awe that leads to activism.

Plus, taxidermy animals only need to be dead once. The fighting bull elephants in the main hall of Chicago's Field Museum were shot in 1905—118 years later they're still there, duking it out and inspiring awe in visitors young and old. The Lincoln Park Zoo, a delightful little free zoo I used to frequent when I lived in Chicago, kept elephants from the early 1900s right through to 2005 when their last elephant passed away, and the zoo decided not to continue their elephant program. How many elephants is that? How many elephants had to live in cramped quarters (elephants can walk 50+ miles a day in the wild,) and spend winters in a place that was famously colder than Mars in 2016, just so people could see them? Confined spaces and climates that differ from native ranges make for a gnarly ethical tightrope when it comes to keeping certain species captive. (At the zoo, I preferred to watch the happy-go-lucky river otters wiggle down their slides and juggle rocks. River otters are known for being playful, but the reason why they're so playful is because they are killing machines. Most animals don't have the time to play too much, they must keep vigilant and look for food. Otters kill like Rambo on a bender so after they fill their tummies, they can just goof off till they want to eat again.) It was the live elephants in the zoo that made me sad, not the preserved ones in the museum.

library by appointment. I started to teach myself taxidermy from books after talking with Joanna about our shared interest in it, and I learned new techniques from a talented artist friend of hers. Soon I was practicing taxidermy on any dead creature I could find. Most car-smooshed rats and pigeons were too far gone for me to use as subjects, but once I started looking more carefully, I found the city was awash in dead birds from window strikes. I started to bring small bags with me everywhere, so I'd be ready if I came across one.

Once, I was on my way to meet some friends for dinner and a movie when I came across a dead dove. I scooped it up, dog poop style, popped it into my backpack and the two of us went on our merry way–I think we saw *Inception*. Another time, I was taking my friend Kate to the Central Park row boats when she pointed out a dead common brown house sparrow lying on the path of the literary walk. This one I put in a paper coffee cup, then into my bag. It all happened about four feet in front of two tourists seated on a bench–I have a small hope the event takes center stage when they recount their trip to New York City. That's usually how it went, an unfortunate critter would be found dead in the street, and, after a brief (and often not so brief) stay in my freezer next to a bag of Trader Joe's dumplings, I'd do my best to give it a second life.

My taxidermy collection has grown over the years to include more attractive, as well as historically relevant, orphans. They make me happy to look at every day. I have other art and sculpture of wildlife around my home too, but there's just no comparing to the natural beauty and scale of the real thing. It would be nice if I could have a living menagerie in my little flat, like when Ace Ventura calls his animal friends out of their apartment hiding places (I can't be the only one who wanted a toilet otter or laundry skunk), but the dead ones don't cost anything to feed, and I don't have to walk them.

A few years after I met Joanna, Morbid Anatomy had grown from blog and library to lectures, classes, and events and was able to open a museum where I had the great pleasure of working. That's where I met Daisy. She was an insect preparer for the American Museum of Natural History. My stories of road-kill rescue pale in comparison to my favorite one of hers.

She was once on a first date when a friend called her with a hot tip about a frozen cat she'd found. Daisy adores cats and asked her date if he'd mind a little detour to pick up the unlucky frosted feline. He was less than supportive of the idea. And thus, the singles-night mixers we hosted at the museum were born. They were an efficient way to gather the sort of people who would respond with positive enthusiasm to the question "do you mind if we pick up a dead cat first?"

The Morbid Anatomy academic blog and museum seek to survey "the intersects of art and medicine, death and culture." And I was lucky to meet a bevy of fascinating authors, scholars, and experts through the related lectures and events we held there. I became even more interested in death culture as it related to our history and modern practices. And the more people I met with a curiosity about death and the past, the more I saw a correlation with a zest for life and an interest in truly being present. The hunters I've met remind me so much more of the academics at the museum than of my coworkers at the outdoor shops. While I could delight in a round table discussion about lightweight backpacking tricks and tips between my hunting and non-hunting outdoor enthusiast friends, I feel like the conversations that might arise about death, history, and culture between my hunting pals and fellow museum nerds would be more introspective and fascinating. I see this similarity between the two in a common reverence for memorial as well.

A portion of my taxidermy collection consists of objects (candleholders, inkwells, trinket boxes) made from horse hoofs. In Victorian-era England, it was commonplace to memorialize your beloved steed in such a way. (Preserving your dog was also a common way of keeping your best friend around forever.) My favorite piece is an inkwell from the famed taxidermy studio Rowland Ward Limited; it has a handsome silver horseshoe and matching cap with detailed silver fetlock hair and is inscribed on top with "Miss Quilp. Died Aug. 25th 1909." Miss Quilp could have been dumped in the ground postmortem like any other person or animal, but instead she was immortalized (part of her), a daily reminder of her to her owner, who loved her enough to bother with such an expense, and a tangible moment in history to me, all these years later. Taxidermy is closest to my heart because of my love for animals, but hair art and a hundred years of death photography have their place in the conversation too.* Mementos of the honored dead, human and non-human.

"They're so beautiful" is a line you often hear from hunters regarding the animals they hunt. It's also referenced by confused non-hunters, unsure how to square the fact someone can kill something they claim is beautiful. This one misunderstanding accounts for most communication issues I see between hunters and non- or anti-hunters. And death is at the root of it. When you become a hunter, you can't help but gain a new perspective on death because you've seen it. The meat buyer and the vegan are also part of an animal's life and death cycle, but they can gloss over the reality of their connection by buying finished products that don't look or sound like animals or that don't reveal the full scope of their impact on the environment. Outsourcing death to someone else allows for emotional divorce and environmental greenwashing. (Ironically, many of my fellow museum enthusiasts have outsourced the deaths of their mounts as well—they adore taxidermy but are suspect of hunting or don't eat meat, an attitude that never ceases to befuddle my hunter friends.) The hunter's acceptance of death and interest in understanding it as a part of life, as opposed to some spooky taboo that's swept under the rug of our culture, is an aspect of the lifestyle that's more actively bonding to the natural world than ignoring it or wishing it didn't exist.

---

*Hair art is what it sounds like. The most common technique was to twist the hair into small cords and coil them around wire to create a framed 3D display. Most often a wreath or bouquet of flowers. Smaller artworks could be made in a locket; larger pieces were more likely to come from the hair of a deceased loved one as they required more hair to complete. Because hair is already dry and dead it makes for an excellent medium if you wish to preserve part of someone.

Death photography was common when the technology was still relatively new and expensive in the 1800s. People who might only be able to afford a few photos in their life would wait for a special occasion like a graduation, military commission, or their wedding. But when they didn't make it that long, as was often the case for young children at the time, a single photo of the departed would become their first and last.

4

The Hunter

My First Deer

This was my eighth time going hunting. All previous outings had been "unsuccessful." I never felt discouraged or like I was owed anything, though. I genuinely enjoyed being in the woods looking at and for animals. But this time, around five o'clock, my luck changed. Two young bucks started walking toward the tree where we perched on our platforms. The sound of my heartbeat was so loud inside my head, I thought the deer must be able to hear it. The first buck wouldn't have been legal for me to take, but the one behind him was a button buck and therefore legally "antlerless." I slowly lifted my crossbow and waited while he got closer. I had what were probably two opportunities before I took my shot, but I quickly found excuses not to take them, rationalizing that I didn't like the quartering angle he was making. Then he got even closer and so perfectly broadside to me that I couldn't come up with any excuse to delay my shot further. I squeezed the trigger.

My guide on this hunt was Fisher Neil, an actor from the Yale School of Drama who stumbled into guiding. "It turned out that once people understood I wasn't the 'trophy hunter' of their nightmares, I became the most interesting subject in the room," he says. Though most of his castmates were young, liberal, city folks, he ended up "really surprised by people's admiration" for his lifestyle. People kept asking him to take them hunting, and he found himself a constant advocate for it. His side hustle as a hunting guide quickly became a well-oiled trade. "After grad school and trying to survive working crappy jobs, it made me think. There were bound to be at least a few hundred people in this huge city with the means and desire to pay for a teaching guide."

His company is called Learn to Hunt NYC, and he says that's how a lot of his urban clientele find him–they just Google that phrase and there he is. The situation is perfect for beginners and people hamstrung by living in the city and not owning a car. If you can get to Jersey City, he can take you from there and back. In addition to hunting trips for deer, turkey, and small game, he provides shooting lessons, basic tracking techniques, and foraging lessons. But the biggest boon to his business has been the apprentice license.

As of 2020, there are forty-seven states that offer some form of the apprentice hunter license. Generally, they mean you can go hunting with an experienced, licensed hunter even if you have yet to take your hunter safety course–like a learner's permit for driving. For Fisher, this means his curious rookie clients can go on a hunt with an experienced guide and see if it's for them before they invest hours of course time in Hunter's Ed or spend hundreds (and what can easily become thousands) on hunting equipment they might not end up using.

Even after his clients get their licenses they sometimes still have him take them out–much easier to go out with a guy who can provide you with everything and save you the hassle of buying much or figuring out how to store it. When just a bow can cost a couple thousand dollars (if you want), it's a pretty good deal to "rent" someone else's equipment and scouting time if you think you'll only get out a couple times a year. He provides a great service for urban-based hunters who are interested in learning but are totally new to hunting, and the apprentice license that makes so much of it possible is an invaluable recruitment and educational tool. I wish more guides would advertise themselves as being beginner friendly when appropriate. It can take the weight off a new hunter who's eager to learn but doesn't know any hunters to approach for mentorship. There are increasingly better resources out there for finding a hunting mentor, but the process can still be trickier for city dwellers.

The first time I met up with Fisher, I had to wake up around 2 to make our 4 a.m. appointment. And even though the trip from my apartment to Jersey City is a simple one, I didn't have any qualms about springing for a cab at that hour (normally I'm just going to bed around 2 a.m. and, upon trying to get up and get dressed at that hour, I barely realized I'd put my underwear on backward, so a car service seemed like a responsible choice, if a more expensive one). This second hunt, I was lucky to have a friend in nearby Hoboken who let me crash at his place while he was out of town, and Fisher was able to pick me up there. We'd be hunting the same area of land we hunted on my first trip. If I pointed to it on a map (and Fisher would kill me if I did), you'd never believe it. It's a relatively small chunk of municipal property surrounded by Levittown-esque suburban housing developments. We arrived in the dark, and I carried his compound bow and the crossbow I'd be using, so he could carry and set up the two tree stands. He made quick work of it, shimmying up the tree about twenty feet like a ring-tailed lemur and setting up one stand behind the other. Then we sat there from 5:30 to about 10:30 am.

You might think sitting in one spot, silently, without moving, for five hours is boring–I can assure you it isn't. Aside from anything else, trying to adjust your body weight so your ass doesn't fall asleep, without moving visibly, is an entertaining endeavor. The little patch of land we sat over was

---

*Antlers are the fastest growing bone material in the world. They also represent the only known regenerative mammal tissue. The annual growth cycle basically goes like this: In late winter, antlers are shed. They fall right off the deer's head, leaving a clean red divot where they used to be. The spot will scab over and heal in a day and new growth will begin just three weeks after they're gone. In spring, velvet, a soft fuzzy skin packed with blood vessels and nerves, will cover and nourish the antlers as they grow. Summer is the most rapid growth period. Older deer will grow their antlers more rapidly than younger ones, and healthy deer mean big antlers—diet, injury, and illness can all affect antler growth. Bumping the velvet or even getting a leg injury while antlers are developing can result in a wonky final product. With so many variables, you can't accurately judge age by antler growth. Peak growth for an animal also depends on the species—could be between five to six years of age for a whitetail, nine to twelve for an elk, or five to eleven for a moose. This can mean two inches per week in a whitetail or an inch a day in an elk. Toward the end of this period, the outermost bone will start to harden. About Septemberish, the antlers have fully hardened, the blood supply stops, and the velvet gets shed. By the fall rut (deer breeding

teeming with wildlife. The sights and sounds of the property were dissonant in the ways you'd expect while hunting land not so far from Brooklyn. I saw five deer, a raccoon, a red fox, more chipmunks than I had ever seen at once, and dozens of squirrels, all while hearing the sounds of the nearby highway. I saw more critters in the morning portion of this hunt than I would all the following November hunting in the Catskill mountains. But that's what happens in densely populated areas, the wildlife gets squeezed together when there's just not that much space for them to spread out. No legal deer came within range for me to shoot though, and by 10:30, the action had slowed down. Deer are crepuscular (active most in the early morning and evening), so we left our stands to have a nap and lunch—no sense hanging around in the tree all day if the deer wouldn't be active again until dusk.

The part of New Jersey where Fisher and I were hunting has a shit-ton of deer (to use a biologically technical phrase). So many, in fact, that state wildlife agents have come up with a clever way of using hunters to help manage the population so it doesn't explode even further. When you buy a deer license for New Jersey, it stipulates that, in the early archery season of particular hunting units (regions), you must shoot an "antlerless" deer first. Antlerless means a doe (who will never grow antlers) or a younger buck that has yet to grow antlers. Bucks will start to grow antlers at one and a half years old.* Before that, they're known as "button bucks" in reference to the

---

period), their antlers are ready to show off to the ladies and strong enough to use for fighting rival males. In mid to late winter, they'll drop them and start all over again. The next time you see a moose swinging around his forty-pound antlers (sixty-plus in parts of Alaska), remember they only took him five months to grow.

What's wild about the process is how we normally don't associate this kind of rapid cell growth with health—because it's basically bone cancer. In 2018, researchers at the Stanford University School of Medicine isolated two genes present in fallow deer responsible for their antler regeneration. Because these genes are also found in people, there's hope further study could lead to treatments for bone issues like osteoporosis. In 2019, a team of scientists from the Northwestern Polytechnical University in Xi'an, China, mapped the genetics of a variety of ruminant species and found evolutionary evidence that deer had tamed tumor and cancerous cell growth to be used to their reproductive benefit, while at the same time becoming five times less likely to develop cancer themselves. The research has exciting implications for future cancer, tumor, and skeletal defect treatments.

pointy bone stubs under their hair that will grow into antlers the following season. Only after you fill that tag can you go after the monster, ten-point, rutting buck you might have been stalking all year through your trail cameras and scouting missions.

Overpopulation of any species (especially our own) is bad news. When you have a species outnumber the carrying capacity of the area it lives in, you end up with sick and starving animals all struggling for the same resources. This leads to two weak deer instead of one strong one, and both are more likely to die. On a large scale, you can see how this would be detrimental to not just a few unlucky animals but to a whole regional population. The antlerless rule keeps the deer population balanced in terms of size and buck-to-doe ratio. Antler restriction laws don't have to be statewide; they can be specific to regions of each state classified as "units."

Each state sets their own hunting regulations. And unlike a lot of other regulations set by states, which may be based off public opinion or a politician's personal beliefs, hunting regulations have historically come from US Fish and Wildlife biologists and ecology experts who spend the year monitoring animal populations and the environmental concerns that may affect those populations and comparing them to the data of the previous years. Science is responsible for determining the best management practices for each local habitat.

The most basic hunting regulations are things like seasons (when you can start and must stop hunting a particular species, including both the time of year and time of day) and bag limits (how many of that species you're allowed to kill in a day or in a season total). The New York State season and bag limit for gray, black, and fox squirrels runs from about 1 September through 28 February. Within that time, you're allowed to shoot six squirrels total per day that you hunt. But let's say, for example, wildlife biologists notice a precipitous drop in squirrel populations due to an extraordinarily cold winter affecting their food sources or something. The following squirrel season could be edited to three squirrels per day rather than the previous season's six. Or maybe a mutant squirrel virus is killing off just fox squirrels—then the fox squirrel might be removed totally from the next season's small game hunting list, and it might remain off the list until that population rebounds and biologists deem it stable again. Turkey season is quite restrictive. The 2021 New York spring season lasted only 1–31 May, shooting hours were "one half hour before sunrise 'til noon," you were only allowed two male turkeys total for the season, and they couldn't be shot on the same day. And the same rules apply: if biologists see a dip in the turkey population, they could decide that the best thing for the turkeys in that region is to restrict the bag limit to one turkey for the season.

In an era of extreme climate change, hunters provide biologists with valuable citizen science data. Because animal population declines are not a result of hunting (because it's so highly regulated and adaptable), biologists can use the information gathered by hunters to understand the most drastic changes taking place in localized segments of the environment. If hunters are reporting fewer turkey numbers one season, that can clue in biologists as to how they should manage the next season and what to keep an eye on environmentally, like increased numbers of small predators, poor-quality forest understory, or a trend in flooding unusual to the area.

I've chatted with non-hunting friends about making my preparations for upcoming hunting seasons, which in some cases can be years in advance of the hunt itself, and been met with total surprise and curiosity. One of my friends had no idea there were any regulations involved at all beyond "buy a hunting license." She thought you just went into the woods anytime anywhere and shot whatever the hell you wanted. As I became more of a hunting advocate, this struck me as one of the bigger communication issues between hunters and non-hunters, most people just don't know how the management systems and regulations are set up.

Fisher and I encountered one such soul when we returned to the woods after lunch. Walking back toward our tree stands from the road, we were passing the backyards of a few houses when Fisher heard some reserved shouting in our direction (I heard it too but paid no attention, as I was used to the indiscriminate yelling of the city and forgot we were in a place where that was less common). He handed me his bow, walked over to where he heard the voice, and found a middle-aged woman telling us we were trespassing on private property. This was not the first time Fisher had heard such a thing, and he went on to have a conversation about the public land and how we were hoping to fill our freezers for the year instead of buying meat from the store. I could hear the tone of her voice change from defensive to something closer to understanding, but I stayed back where I was behind the trees, not wanting to crowd her, especially not while holding a Whitman's Sampler of archery equipment. As they finished talking, I heard her say "Well, I won't wish you luck, but have a good day." I thought this was a positive exchange. She'd had a personal interaction with a friendly well-spoken hunter, she'd learned about the public land right behind her house, and she'd been reminded that people eat what they hunt, a fact I think a lot of people forget when they're furiously replaying the storyline of the blood-lusting, trophy-seeking, animal murderer in their heads. I told Fisher I wouldn't have been able to stop myself from asking if she was a vegetarian. We both had the sneaking suspicion that she was not.

In a 2017 article for *Outside*, Wes Siler writes about hunting's poor reputation, pointing out that "Something that contributes huge amounts of money to conservation is seen as a blood sport. The most humane way to put meat on your table is seen as cruel." *Outside* magazine is generally geared toward non-hunting outdoor enthusiasts—as someone who spends a lot of time talking about hunting, taxidermy, and conservation with non-hunters, I'm always curious how messages like Siler's are received. It's strange for me to look at his words now and understand them. When I was a young member of the "blood sport and cruelty" camp, I was acutely aware of the fact that many of my conservation heroes, especially Theodore Roosevelt, were avid hunters, but I dismissed this as an outdated anomaly, a necessary evil for putting food on the table or creating museums that would be responsible for the education and inspiration of future generations of conservationists like me. Much of this perception was based on the ways hunting was depicted in popular culture when I was growing up.

Most people I talk to who are my age or older can trace the vilification of hunters in their minds back to *Bambi*. No surprise there. I felt the same way when I was a kid. The film had such an impact that it has a phenomenon named after it, "The Bambi Effect," which describes a public prone to anthropomorphism,* particularly of animals considered "cute," at the expense of realism.

Perhaps the most aggravating aspect of the whole Bambi thing is that the book the film is based on was meant to be an honest representation of how brutal nature really is. The story is full of ghastly, slow and painful deaths inflicted by predators and same-species rivalries. The author, Felix Salten, was a nature-loving hunter himself and detested poachers (like most hunters do), so Man, the villain, still plays a role, but even without Him around, life in the woods still hinges on death. Salten's point was that we humans might foster a more sustainable relationship with the outdoors the sooner we come to terms with that. His writing was an attempt to draw attention to the Madonna-whore complex people had developed toward nature–ironically, the very issue the film version would be most responsible for proliferating.

In *Call of the Mild*, Lily Raff McCaulou astutely identifies the men in *Bambi* as poachers, not hunters. They use illegal tactics (setting forest fires to drive animals in a desired direction) and hunt out of season (Bambi's mom is shot in spring, but deer hunting season is in fall). Does weren't considered "fair game" at all when the film was made–famed conservationist Aldo Leopold was lambasted when he suggested opening a season for does in an area of Wisconsin to control the exploding deer population. This was a time when deer populations were still in recovery in parts of the United States, and the science of game management, which Leopold basically invented, was still in its infancy. Now, it's common to find tags for females and young males in various recovered populations (like the unit in New Jersey where I was hunting). Unfortunately, you'd have to know something about hunting to pick up on those subtleties (it might have been slightly

---

*Is anthropomorphism good or bad? Yes.

Giving animals human traits is the nuance of all nuances. Animals communicate, and some species are easier for humans to talk to than others. On the domesticated side, dogs are the easiest. Dogs are social predators, like us. Cats are predators, but they're solitary. Horses are social, but they're prey. You can say "my dog is happy" and probably be right, and it's no big deal. It's when humans impose human social and family structures onto other animals that we enter dangerous territory, because those species have their own social structures that need to be understood in order to help them best when they need it. Kind of like how the American hand symbol for okay means "asshole" in a bunch of other countries. Before you assume the intent or desires of any animal, study that species like a wildlife biologist and learn its language. Otherwise, you might end up calling it an asshole.

more common knowledge for audiences around the time of the film's 1942 release and the hunting boom that followed the end of World War II, but it certainly isn't now).

Adam Mullins-Khatib is a film critic and producer. I asked him to weigh in on the Bambi Effect and if he thought seeing more hero hunters (or at least hunters who aren't villains) in movies and television would be enough to change public perception. He agreed that the use of illegal or frowned-upon practices by movie poachers, as in Bambi, probably doesn't register with most people because, "hunting itself is perceived by outsiders as not really having balanced rules in the first place."

> *We all inherently know on some level that the tactics "evil" poachers use are "unfair," but beyond that, to the average person, so is the idea of legal hunting. One of the most common exclamations against hunting I hear is the horror that defenseless creatures are killed for sport. If we're starting from that premise, I have heavy doubts that the nuance and technicalities of hunting versus poaching is really playing that much in our viewers' thought process.*
>
> *We live in a contemporary America where the very idea of hunting is coded as conservative and rural, and with an increasing portion of our population living in urban areas, it's tough to see a clear path for expansion of understanding of hunting practices to larger and larger portions of the public. I come from a family that lives in a rural area and dabbles in hunting and fishing, while, at the same time, I personally live in a massive urban center where the closest I come to nature on a regular basis is a city park. If, I the urbanite, didn't interact with people who hunt regularly, how would I ever even get the chance to understand hunting? And if I don't know the basic practices and benefits of hunting, I'm certainly not going to be able to view passing images coded as hunting or hunter with any sort of understanding beyond my preconceived notions of what people who hunt are like.*
>
> *Hunting is seen as an esoteric activity, requiring either equipment that a large swath of the population doesn't understand and is terrified of, or a mystical sense of the natural world that can only be gained from spending countless hours in the wild, sacrificing the amenities of our modern lives. There are seemingly indecipherable symbols, practices, and techniques in the culture of hunting, which don't feel accessible to an outsider. Part of this dynamic includes idolization of a lost past in which larger segments of the population hunted, but,*

like with much romanticization, it's a preference toward nostalgia without a true desire to go back to that place. Seeing hunting in movies reminds us of that past, with all the cultural and historical baggage that comes along with it. It's a past that feels simultaneously familiar and uncomfortable, tying us to a natural world that is being filtered and presented to us through the big (or these days very small) screen, in our dark, climate-controlled theaters or homes, about as far outside of that world of nature as we can be.

That said, I do think that an increase in "normal" people hunting in films—not heroes, not villains—would have at least some positive effects on our feelings toward hunters

My friend Stephen Sajdak is one of the four cinephiles who make up the popular (and very funny) "bad movie podcast" *We Hate Movies.* I asked if they, too, would touch on this hunting-in-movies stuff, and, between *Simpsons* references, the gang was eager to bring up some issues I'd missed and look at my critiques not only as film theory buffs but from the perspective of non-hunters as well. Steve was quick to point out how divorced fishing is from hunting in film—how it always communicates something gentle, calm, and pensive, as opposed to hunting, which often finds itself mixed up with violence or toxic masculinity—everyone agreed guns and blood have a lot to do with the difference as perceived by an audience. Political perceptions and coding dominated the conversation.

In movies, a set is rarely dressed with any trappings of the hunting lifestyle without it meaning something so important that it becomes less of a personality indicator for the character and more of a major plot point. An audience can see a mountain bike in the garage, a badminton racquet sticking out of a bag, or a set of golf clubs in the trunk, and pick up that a character is sporty or wealthy without needing to see them playing badminton or golf. But show anything related to hunting, and the fact that the character is a hunter must come up again. A real Chekhov's gun situation. Exceptions to this rule come when someone's conservativism is central to their character. In Mike Nichols' (perfect) ensemble comedy, *The Birdcage*, the camera can show a handsome wood and glass gun cabinet in Gene Hackman's study because we know his character to be the cofounder of the ultra-conservative political action group the Coalition for Moral Order. We as viewers already associate gun ownership and hunting with conservatism so no further explanation is needed. But what if we saw the same cabinet in the apartment shared by the gay couple played by Robin Williams and Nathan Lane? I bet that would leave us with a lot of questions. Hunting has become such a touchy and politicized topic it can be difficult to represent it casually as a part of someone's lifestyle, especially if that person exists outside the stereotypes that were constructed for them long ago. A box of rifle rounds sitting on a work bench will always carry more weight than a few scattered shuttlecocks.

Characters in certain settings are sometimes shown hunting so the audience can surmise how patient, clever, and practiced they are. Think hero hunters like Katniss Everdeen, the arrow-slinging protagonist of *The Hunger Games* books and film series, or Merida from *Brave*, a Pixar film about a Scottish princess who is a skilled archer. Katniss has been credited with the surge in archery's popularity since the 2012 release of the first *Hunger Games* movie, and Cameron Hanes, a sponsored bow hunter, and Grizzly Jim Kent, an expert traditional archer, have both called *Brave* one of the best representations of archery in film (which makes sense when you consider how much easier it must be to animate perfect technique than to teach it to real human actors). However, *The Hunger Games* is set in a dystopian, alternate reality (no grocery store), and *Brave* is a period piece that takes place in a romanticized past. In modern-day, real-world settings, there's always a caveat for the hero hunter. *Captain Fantastic*, from 2016, follows an endearing family living a rustic lifestyle and enjoying hunting as a part of their self-sufficiency and family culture. But because they are so removed from society (they're off the grid, the kids are home schooled, they rarely venture into town, no grocery store) hunting once again becomes a vital necessity as opposed to a lifestyle choice. In 2017, the film *Wind River* depicted a positive representation of a hunter who uses his tracking skills to help solve a murder. But he's not just a hunter, he's a US Fish and Wildlife officer, one who spends the opening portion of the film recording the account of a rancher losing livestock to a local mountain lion. Our hero doesn't want to hunt the mountain lion–it's his job. In movies, heroes hunt because they have to–villains hunt because they want to.

Of all the famous pop culture hunters, from Elmer Fudd to Ben Gazzara's one-man mafia operation in *Roadhouse*, the depiction that bothers me the most is in the seemingly innocuous *Harry and the Hendersons.* In the film, John Lithgow plays George Henderson, an accomplished hunter who gives up hunting after meeting Harry, a legendary North American forest ape or sasquatch. In the first half hour of the movie, Harry is disturbed by a mule deer shoulder mount in the Henderson house, so he takes it down and buries it in the backyard. Immediately after, he eats the family's goldfish. Again, here's that perception that it's okay to kill animals for food, but *hunting*, well that must just be for fun, right? If Harry only knew Mr. Henderson's license and tag fees were paying to keep the woods of the Pacific Northwest a suitable habitat for him and the other woodland creatures that grace the screen while the opening credits roll, maybe he wouldn't have been so hard on George's taxidermy.

Hunters use a lot of different words when they talk about hunting, shooting, or killing an animal. I used to see this as misdirection, a calculated ambiguity to avoid what I figured hunters must have known was the true ugliness of their actions. There is something more gentle, softer, about these words, but today, I recognize them as part of the human tradition of using euphemisms. What's so different about a hunter saying *taken* instead of *killed* and a grieving person saying *passed away* instead of *died*? Death is still taboo, despite the fact it comes for us all, but beyond that, isn't it more respectful? To the deer and to your grandmother?

Hunting euphemisms are couched in the experience of hunting—what it adds up to at the end of the day and means to the hunter. Seeing these words from the other side, I find myself using them too, and in the spirit I now know they're intended.

Harvest To kill an animal you plan on eating. This is probably the most common term used by hunters to describe their kill. I had a real problem with this one before I started hunting. A farmer plants some seeds in a plot then, after tending to those seeds, can harvest a nice plump pumpkin. Are you really *harvesting* that deer? Are you responsible for raising it up? Well, as a hunter in the United States, you kind of are. Look at it as if the backcountry of US public land is the plot you're responsible for, and the carefully managed game, which belong to all of us, are the crop. You've paid for the land to be protected, the water to be clean, and the animal populations to be monitored and aided. That's where your license fees, equipment taxes, and game tag costs go (except, no matter how much you pay you're never guaranteed to come home with anything). Tennant number one of the North American Model of Conservation says wildlife is a public resource. This goes the same for people who enjoy foraging for wild greens, mushrooms, ramps, or fiddleheads. The forager enjoys our collective natural resources, only they didn't pay to protect the land and water. There is no program to ensure all the ramps taken from the land regrow the following spring. If a modestly priced foraging license was added to state regulations right next to fishing and hunting licenses, that could be an effective way to ensure people were educated about the safety of wild mushrooms, and how to harvest particular wild plants without disturbing their ability to grow back the next year. What is hunting if not foraging for meat?

Taken Shot or killed by or with. Taken is just another way to say killed. That deer was taken with a compound bow. That bull elk was taken by American spy, socialite, and badass Gertrude Sanford Legendre in 1920. This word is the least meaningful euphemism for kill. Softer and more respectful, but with less motivation and meaning than harvest.

Connected With Shot or killed. Same as taken. The connection part refers to the physical connection of a bullet or arrow leaving your gun or bow and literally connecting with the body of the animal you're hunting. At the same time, it also encompasses the entirety of the hunting experience by referring to the exciting and emotional climax of a "successful" hunt (if you want to define coming home with your quarry as the pinnacle of success). I was in the woods for at least seventeen days during the hunting season one fall, often twice a day, before I connected with a buck on my first successful solo hunt. *Connected with* seemed more appropriate for me than *harvest* or *taken* in this instance. Sure, I harvested the meat, and I took the deer with my friend's Weatherby rifle. But after so much time and work, from the first day I started Hunter's Ed to the emotional experience of killing an animal for food by myself for the first time, the idea of *connection* made it by far the most fitting term for what had finally happened.

Chasing A playful replacement for hunting. As in "I spent winter chasing rabbits in the mountains." It makes me think of the part in Jaws after Quint gets a barrel into old Bruce (Bruce was the animatronic playing the shark). The music takes a dramatic tonal shift from a scary and ominous lower register to a bright and exciting higher one. I think we're programed to think of hunting as serious, and though hunters take their shots very seriously, there's no reason you shouldn't admit to the hunting part being exciting. Frolicking in the woods with your fellow predators and prey, just another animal in nature—dare I say it's fun? And while *hunting* still just means to pursue an animal, *chasing* serves as an even better reminder that often, in fact, most of the time, it's just that, a chase, and you are more likely to return home empty handed than you are with dinner. Those seventeen days I spent in the mountains absolutely felt more like a chase, and the deer were always a few steps ahead of me.

Catch A replacement for kill, *catch* is used exclusively by non-hunters. I'd seen memes online of hunters having fun with non-hunters using the term *catch* to refer to the critters they're hunting, but I didn't quite believe it until I heard the same thing issue from the mouths of my own friends. "Did you catch anything?" they'd ask, and I'd instantly picture myself with a tall Elmer Fudd hat, running around the woods with a giant butterfly net. I've never met a hunter who has a problem with this, but it's generally regarded as pretty silly. However, in the interest of ambassadorship and creating hunting allies, it's worth analyzing why non-hunters use this term. While a non-hunter can obviously surmise what *harvest* means, it's perhaps not a word they use or hear often in the context of hunting because, really, how often are non-hunters actually talking about hunting? So, if they don't know *harvest* is a hunting term, and they don't want to use *kill* or *shoot* because it seems too graphic or somehow impolite, then *catch* starts to make a lot of sense. It's what people say when they go fishing and how different can hunting be really?*

This hunt with Fisher was the first time I had connected with a deer, and I felt shock more than anything else. When we got down from the tree, it was easy to find my bolt and the start of the blood trail. There was blood on the tree behind where he'd been standing. I looked at the blood on the arrow fletching and on the surrounding leaves and could not believe how bright it was. It was so bright I thought if you saw the same color blood in a movie, people would think it looked fake–it was like a red highlighter. The brightness was indicative of freshly oxygenated arterial blood. Unlike the maroon, low-oxygen blood from a vein (the kind you'd see if you donated to the Red Cross). I was comforted by this evidence of a heart shot.

As I started following the red-spotted leaves, Fisher was mindful to give me this moment to walk ahead and follow the blood trail at my own pace. I could see the path the deer had cut through thick bushes that had been suddenly parted, and when I reached the other side, I saw him lying dead under a tree. We weren't far from the tree stand, under forty yards, I think. The first thing I felt was relief. I was so relieved I had made a good shot and that he had died quickly. I was relieved it was over, that I had done it. I had been so sure I was going to burst into tears when I finally laid my hands on him, but I didn't. I was more in awe, and still most certainly shocked. I felt his ear and stroked him from the top of his neck over his shoulder to his soft white belly. I want to tell you he was a grand majestic beast, but he was rather scroungy, thin, with patchy hair and big warts near his eyes. (Deer "warts" are benign tumors, cutaneous fibromas. They're ugly, but harmless to people and mostly just annoying to the deer unless they get in the way of their vision.) He was beautiful to me though. I felt proud, like some Bizarro World reverse parent. It was strange and powerful.

---

*One word that seems to permeate hunting culture from both the inside and out is *gamey*. Here are the appropriate uses for gamey, to be used by hunters or non-hunters: 1) You have never eaten any meat other than beef, pork, or chicken and therefore have no words to describe a meat flavor that is unlike those three. Try using any other word. Pretend you're describing a fine wine and get creative. 2) You fucked up dinner and are trying to cover for it.

It's easy to talk about hunting and our primal selves because the bodies we occupy today were built by and for that very purpose over the last two million years. The reason we're so "smart" is because meat gave us a more calorie-dense food that required less work from our guts, and we used that saved bodily effort to develop our brains. (Which are so big at this point that human babies have to be born half-baked, helpless, dummies. Any further development and their giant heads wouldn't be able to fit through the birth canal.) After a book reading at Greenlight Bookstore, Steven Rinella found himself discussing this topic with a vegan who came to the launch of his memoir *Meat Eater*. Rinella makes the case that, in regard to our evolution from early humans to modern-day people on this earth, it's way more bizarre to *not* be a hunter. He makes other compelling arguments as well, but I particularly enjoy the reminder of what a blip we are in the evolutionary timeline. It's important to be careful not to use that timeline as an excuse for modern choices, but there is understanding to be gained from it. You can't always think your way out of something you feel. (For more on this read *Deep Survival* by Laurence Gonzales).

I know the difference between various venomous and nonvenomous snake species, but I still jump when I'm surprised by any snake in the woods. That doesn't happen to me with chipmunks. The reason snakes and spiders, dark places, murky water, and sleeping in a new place makes us uneasy is because those things are supposed to. We may not know if a snake or spider is venomous, so it's better to stay clear. Our senses of smell and hearing are pretty pathetic, so we often rely most on sight, which makes dark places scary. This sets up an outright annoying battle between our modern minds and the two million years of hunting and meat-eating evolution that gave us those minds. For me, the idea of our primal nature buried within our modern selves was always a jaw-tightening, eye-rolling point of contention. Until I took part in an activity that revealed it and I actually felt the exhilaration and reward for myself, despite simultaneously loving animals and not believing I could be capable of the act of killing one. In my head, I never would have thought my heart or gut would have responded so positively. I haven't always been able to describe how I can love animals and hunt them at the same time, but I now see it as equal parts heart, head, and gut. And I can only see it that way because I've had the chance to feel it.

Housing the evolutionary hardwiring of a predator and the mind of a modern civilized person in the same body comes with conflict. I'd argue that kind of conflict is good for us, though it doesn't make things easy. Rinella, a lifelong hunter, still feels it himself. While considering the gravity of a successful hunt in *American Buffalo*, he says, "I feel compelled to question what I've done, to compare the merits of its life with the merits of my own. . . . It's more complicated than guilt. . . . I feel an amalgamation of many things: thankfulness for the meat, an appreciation for the species, and, yes, a touch of guilt. Any one of those feelings would be a passing sensation, but together they make me feel emotionally swollen. The swelling is tender, a little bit painful. This is the curse of the human predator, I think. When a long-tailed weasel snakes its way into a rabbit's den and devours the blind and hairless young, it doesn't have to think or feel a thing. Watching a weasel, you get the sense that a complete lack of morality is the only path to moral clarity."

Rinella never really made the choice to become a hunter–he was just born into a family where that was the norm, but someone like Lilly Raff McCaulou, who made the choice to start hunting later in life, after careful considerations on behalf of the environment and animal welfare, still struggles with the same feelings. McCaulou describes herself as seeing animals with "a whole slew of anthropomorphic traits," and details the consequences of such a view. "This strikes me as the environmentalist, vegetarian, animal lover's approach: Any death of any individual is painful and bad. The trouble is, I now think of animals both as members of a population *and* as individuals. It makes for a lot of handwringing. But maybe it's a necessary paradox; it's what makes me a responsible hunter."

This admission of paradox rang true for me. I am an environmentalist and an animal lover. But the more I hunt, the more it all makes sense. Not because I'm not bothered by the death of an animal, but because, more and more, I get to see myself *in* the cycle of nature as opposed to outside of it. Rinella likes to call us human predators, but I prefer human animal. Hopefully it sounds redundant to you. While we might forget that we are predators, many of us have forgotten that we are even animals, and I think that's an important place to start as we try to find our place in a world filled with animals, human and otherwise.

In the past few years, Fisher has played a part in more first-time hunter successes than just about anyone. He says that everyone responds differently, "Some people are emotionless. Some feel terrible. Some are incredibly happy. A few become ill when we field dress the deer. Most are shocked that it has actually happened." Standing and looking at my button buck, I was suddenly conscious of Fisher beside me and the fact I had to gut the deer and get him out of the woods. I dragged him out from under the tree by his two front legs and Fisher set up his camera to take a picture. I picked leaves to cover the wound from the bolt and to put in his mouth—a Bavarian tradition meant to give the deer his last meal before you, in turn, consume him—and I realized that was respect. I wanted to show respect, even though he was dead and didn't care. I cared. All those cliche lines about the relationship to the animal you've just killed became real (except "sacrifice" ugh, please never say sacrifice—we are not gods).

Author Timothy Ferriss described himself as an anti-hunter in his 640-page tome dedicated to learning new things, *The 4-Hour Chef*—then he proceeded to have a transformative experience while gutting his first deer in the woods. "Before scheduling my hunt, I vowed not to romanticize it. I had no desire to be Hemingway, nor glamorize killing. But something very odd did happen: thirty seconds into gutting the deer, once my hands were inside and heated, it felt like second nature. The anxiety vanished, and I was inexplicably good at it. How do any animals in isolation know how to behave, how to hunt, how to do anything? Hardwiring, I suppose. I've never experienced anything like it."

Robert Rockwell, one of my taxidermy and museum heroes, expressed similar surprise in his 1955 memoir *My Way of Becoming a Hunter*, when he reminisced about his earliest days in the field saying "Was it chance or instinct that made me first look on hunting as the most gratifying thing in the world? It satisfied curiosity and the inquisitive desire to come to close quarters with what was beyond restricted horizons. Possibly it was that primitive instinct for the hunt which dates back to the dawn of all created things. Some of us never suspect the existence of such an impulse. In most of us it is smothered by our artificial way of life and the mad race for comfort and security."

Maybe you've never flinched at a snake—I can't tell if that means you're more or less evolved.

Taking a picture with your kill is colloquially referred to as a "grip 'n grin," the same phrase we use when describing the cheesy handshake and smile seen at political events or when someone is receiving a diploma. There's been a surprising amount written by hunters and anglers about the grip 'n grin. I feel like I see a new article about the practice and our human desire to share every time Instagram refreshes or I'm sent the latest newsletter from *Outdoor Life*, *Field & Stream*, *Blood Origins*, *Hunt to Eat*, *Backcountry Hunters and Anglers*, *Meat Eater*, or *Orvis* to name a few. It's perhaps an even more complicated topic today than hunting itself, because of social media. Why do we post anything for other people to see? Pride usually, and that's not necessarily a bad thing. It's okay to take a photo of your dinner if you're proud of the work that you put into making it—so why isn't it okay to take a photo with an animal you successfully hunted? Hunters are proud of the source of their food, not just the cooking of it. And they should be. A successful hunt arguably took more time and work to achieve. Hunters want to share their story and their success, but these photos can spark anger in anti-hunters—or worse, there's some jackass out there showing off what we would all agree is a disrespectful photo that manages to give every hunter a bad name. The conversation now has splintered into multiple questions nobody in the hunting community seems to have an answer to. Should hunters stop sharing their hunt photos? If thoughtful hunters stop sharing so as not to accidentally offend anyone, what about the assholes who don't care about offending anyone? Positive role models need to be visible. How else to normalize sourcing wild foods?

I wasn't sure what to do in my grip 'n grin. I purposefully hadn't thought about it ahead of time in an effort to not jinx myself, not to secretly become cocky, and not to expend mental energy on something I wasn't sure I was socially "allowed" to do as a modern hunter. The Academy of Motion Picture Arts and Sciences will tell you to always write your acceptance speech, however. And every deer, even my mangy, wart-riddled button buck, is a prize—an award, a reward, for having done things correctly. I knelt behind him and held his head up with a blank stare, thinking "is this respectful enough?" But it didn't feel right, and I couldn't stop myself from grinning when Fisher snapped the third shot.

As soon as our little photo shoot was over, I filled out my deer tag and wrapped it to his front hoof. In addition to the physical tag, I also used an app on my phone to "report my harvest" to the New Jersey Department of Fish and Wildlife (reporting what you kill lets Fish and Wildlife compare license sale numbers to successful hunters). Then we prepared to start gutting. Gutting is generally done in the field as soon as possible. Especially if it's hot out. When an animal dies, gasses start to build up inside the digestive system and the whole process goes a little easier the sooner you can start. Removing the organs also gets the bacteria-filled gut outta there and helps the body cool faster, which aids in preserving the meat for the time it takes you to process it or take it to a butcher.

I'm not squeamish, so the act of cutting the deer open from sternum to pelvis wasn't emotionally difficult for me, but I was being excessively slow and cautious not to cut too deep or puncture anything I shouldn't. I asked Fisher if I was doing everything correctly and if whatever I had my hand on was the right thing for me to have my hand on. I felt like a rookie bomb squad tech, as if I could do something so incorrectly the deer would just explode in our faces. I wanted to do such a good job. I would have been devastated if I managed to do something that would have led to any meat loss.

There aren't many steps to gutting–I'd say between six and ten, depending on your preferred methods or what you deem a "step." I didn't just want my first gutting in the field to go smoothly, I wanted to clearly see and feel each step so I could remember it for the future when I was on my own.*

Fisher pulled at the heart, cut it free, and handed it to me. I was relieved once again–now I had proof I'd made a heart shot–the entry and exit points were clearly visible. I knew I probably wouldn't make a heart shot every time I went hunting, but I was so happy I did this first time. I was also

---

*Even with my careful rehearsal, one deer I eventually took on my own would end up victim to a mistake I made gutting—I nicked his intestines without realizing it and made a mess when I went to pull them all out. My buddy had to help me rinse the green flecks of partially digested vegetation out of the cavity, while I held his whole body up with my arms wrapped around him under his front legs. I'm grateful my first experience gutting went well.

just a little bummed about the damage, as heart is one of my favorite bits of meat. I slipped it into a Ziploc bag and put it in my backpack. Then I severed the deer's windpipe and used it to pull the remaining organs free from the body cavity. They flopped out onto the ground in one big, wet, pink pile. After taking the heart, liver, kidneys, caul fat, and whatever loose bits you wish to preserve, you just leave the guts. Various woodland critters will quite happily clean up the rest after you.

As we were finishing up, Fisher saw a deer in the distance, a larger buck. I couldn't see it and didn't want to spoil his opportunity by bobbing around looking for it, so I just crouched down behind some bushes next to my deer while he positioned himself for a potential shot. I found myself unexpectedly thankful for this distraction. Since we had first reached my buck, I hadn't had a moment to just sit and appreciate him. I'm sure Fisher would have given me all the time I wanted, but I was too caught up in the moment and my own shock to stop and take that time. Now I found myself petting and inspecting every inch of him with wide eyes. Fisher turned back to me a couple of times to fill me in on what he was seeing the other deer do, but I was totally zoned out in some sort of post-adrenaline high, communing with the dead. I'm not sure if I played it off or not.

The other deer moved on, and Fisher came back to me and told me to find a short, sturdy stick, unpacking a length of paracord that we could fashion into a deer drag. He looped one end around the deer's neck and the other to the center of the stick, so I'd have a wide handle to hold onto while pulling him through the woods. When Fisher went back to our tree to break down the tree stands and pack up our stuff, I had another moment to appreciate the deer, the weight of him, and the work it took to drag his body over logs through the woods. I was glad it wasn't easy. I still felt, despite all the empty-handed hunts I had been on, like I hadn't earned my dinner just yet. I took a break to readjust my grip on the stick and stretch out my hands, and when I did, I thought he looked so pretty laying in the bright green grass I took another picture of him.

Fisher carried the tree stands in his backpack, and the two bows, while I dragged the deer. We switched a couple times on the way back to the car. Then we loaded the deer into a metal basket hitched to the back of his SUV and started out for the butcher. I would have liked to process the deer myself, but it was already 7:45 p.m. and would have been far too complicated without my own car, a vacuum sealer, and a place to put an entire deer, so I was content to have someone else do the work for now. I considered killing the deer to be enough physical and emotional labor for my first go around.

The butchers who process deer aren't like the shop where I worked. First, it's a seasonal gig, generally going from mid-September through mid-February. The licensing is a bit different because they're handling wild game and not farm animals, and the same way you might think of a food safety inspector dropping by a restaurant, a deer processor is just as likely to have a conservation officer (game warden) drop by to check hunter licenses and tag information and compare it to the deer being brought in. Some butchers will give you an array of cuts to choose from, but options might be limited if they need to save time (they might be handling a lot of deer all at once if it's the start of archery or rifle season). A butcher I know in Rockland County New York has his own mobile shop so he can come to you and process your deer in a couple hours when you're ready.

Ideally you want your deer to "hang" for a few days. That could be outside if the temperature is cool enough, or in a proper meat locker. The time between death and butchering helps the meat relax and encourages natural enzymes to start breaking down connective tissue. Older deer should hang for longer, owing to the fact they have more of that tougher tissue. It's not much different than aging beef. Mississippi State University's Deer Ecology and Management Lab recommends up to eighteen days of aging if you have access to a controlled-temperature environment, the sweet spot being the mid-30s (Fahrenheit), just above freezing. On the day we brought my deer to the butcher, he joined at least twelve others who had come in before us–lined up on the ground like furry, brown, headless Rockettes–plus one or two more that were already hanging, ready to be skinned.

If you want to mount your deer and aren't sure how to skin it properly, a deer processor can usually help for a fee. Taxidermists need animals to be skinned in specific fashions depending on the type of mount–a shoulder mount you might traditionally see on a wall requires all the hide from behind the shoulders forward (the more the better), while a full-body mount requires everything you can see from the outside. Knowing how to prepare an animal for the taxidermist can make a big difference in the quality of your finished piece. But as much as I love taxidermy, I wasn't going to have him mounted–keeping his skull would suffice. I still get to see him every day.

My deer earned me a little over thirty-six pounds of boneless meat: seventeen of ground venison; sixteen of roasts and loin; three of kielbasa, hot dogs, and landjager. I know this because, once I got it all home, I made my friend Matt come over with his kitchen scale and weigh every vacuum-sealed piece. (Fisher was kind enough to get my meat from the butcher the week after our hunt so I could just swing by his apartment to pick it up. I brought it home on the subway in two tote bags and was glad I had cleared enough freezer space for it all. Now, after bringing my old Jeep out from my dad's place in California, I do all my own meat runs.) The butchering cost me $113 (that's $15 more than the base price, because I *had* to have sausages and those cost extra), which brings the cost of processing to $3.13 per pound. Add the cost of my out-of-state license and guide, and the price jumps to $17 per pound which is still a deal for local, eco-friendly meat (don't say organic–we have no idea what my deer was eating in the backyards of New Jersey–when you get your Brooks Range Alaska Dall sheep tag, then you can probably say organic). Factory-farm meat at the grocery store will be less expensive at the counter than my venison (or than meat from a small sustainable farm), but it costs a great deal more than what comes out of your wallet at checkout. The cost is borne by the animals who have to live on industrial feed lots, the people who work those lots, the people who grow the crops on the monoculture farms that make the food for the feed-lot animals and the plant-based meat substitutes, the planet that has to put up with it all, and all of us who have to live here.

The cost of meat from hunting changes over time. Your first few seasons, you'll have greater up-front equipment costs. Maybe you have the space to do your own meat processing, which means buying a couple good knives, a good vacuum sealer, and a meat grinder, plus more of your time, but no more paying the butcher. Maybe you start hunting with a guide and then switch to hunting solo. Then there's the license and potential extra tag fees if you're out of state. My first deer cost me a hell of a lot more than my second. That's because the first deer was out of state (so the license cost more) and I had paid for a guide. My second deer was in my home state, and I was solo. I had a butcher for both, but with the difference in license cost and going it alone, the "price" of my venison dropped by 82 percent. Bringing the cost of my local, "wild caught" specialty meat to just over $3 per pound (I ended up with a little more than forty pounds of boneless venison from my second deer). When you consider that the average retail cost of grass-fed beef filet is $28.98 per pound (as listed in the January 2021 USDA National Monthly Grass Fed Beef Report), my venison is quite the bargain! It took a lot more of my time to acquire, but if you normally only buy a couple pounds of meat per grocery run, that's twenty visits to the market where I didn't have to buy meat. And for me, the time in the woods is always worth it. Most new hunters aren't taking up hunting with saving money on meat as their priority though. I did it to know where my meat was coming from and where the money I spent to get it was going.

After the butcher, Fisher and I drove back to Jersey City chatting about duck-hunting strategies, the New York theater scene, hunting abroad, and favorite recipes. He dropped me at the train station around 9:40 p.m., and I thanked him again before he drove off. Now I found myself dressed in camo with a smattering of bloodstains on my hands and a heart and severed head in my backpack, boarding the PATH train then taking the subway through Manhattan and Brooklyn. I wasn't too self-conscious though; I probably wasn't even on the top ten list of bizarre things people had seen on the train that day. The ride home proved to be the perfect quiet moment of reflection, as so many evening subway rides are. I sat there just staring at my backpack resting between my feet and looking at the photos I had taken as if they were from years ago. Fisher had taken a video from the moment he saw the two deer coming closer to our path until my buck disappeared into the bushes after being shot. I watched it for the first time with the same heart palpitations I'd had sitting in the tree stand.

I'd worried I might be a complete wreck after killing my first deer, but even though I did (and still do sometimes) have complicated feelings about it, which included a sort of apologetic sadness for the demise of my particular deer, I felt good. I felt even better the following night when I cooked up the heart with butter, garlic, and shallots. It was the first meat I had ever eaten that was environmentally and emotionally guilt free.

5

Pigs / Hogs / Boar

Invasive Species and the Model

I was having trouble slipping the little .22 LR cartridge into the rifle. They're tiny and my hands were sweaty. It didn't have anything to do with handling a gun—I just have excessively sweaty hands, a condition called palmar hyperhidrosis. Occasionally, I get Botox injected into my palms to stop the sweating for a few months. The procedure (needles jabbed repeatedly into your fingertips) is torture-scene-in-a-period-war-film painful. (There are dozens of us! Two to three percent of Americans suffer from excessive sweating of the palms and feet.) I felt the need to explain the entirety of my unusual medical condition to my companion, Chris, so he didn't have any qualms about my comfort level with the smallest of the three rifles he'd brought to the range that day.

I had only met Chris a few months prior, at a delightfully nerdy tech conference in Chicago where I was speaking about the history of taxidermy, and he gave a talk on the terroir of Texas wild meat. I chatted him up after his talk and told him about my cautious interest in hunting. Then I flat out asked him if he'd take me on a wild hog hunt if I flew to Texas. He flashed a polite but warm smile and generously gave me his email.

If I say "let's get coffee sometime" I actually mean it. Likewise, if I invite myself over to your Texas home for a feral pig hunting trip and you say "sure," you will be hearing from me. So, my first email to Chris was basically a warning, giving him one more opportunity to escape the burden of a houseguest. I learned later that hunters make it a point to mentor rookies–and Chris often coached newbs like me. Aside from the fact that it's just fun to share something you love with someone new, if you want to learn to hunt as an adult, there aren't many widely available schools or camps to teach you. There certainly weren't when I first took up hunting, though they are becoming more common. A few new ones have sprung up while I've been working on this book, mostly run by larger organizations, but I expect to see an increase in beginner-friendly outfitters like Fisher too. And though I've seen a tremendous change in the availability of resources for new hunters, they can be tricky to find if you aren't sure where to start looking.

This was all long before I connected with my first deer. It would be my first hunting trip, and I was thankful that it would be for feral pigs (which can be called boar, pigs, hogs, swine, and wild or feral in any combination). The fact that they're invasive, overpopulated, and an ecological nightmare for native plants and animals helped me make peace a little faster with the fact that I was there to kill one. *Smithsonian Magazine* referred to feral pigs as "a plague" in 2011, and the *Atlantic* called their spread a "feral swine bomb" in 2020. Plus, I eat a lot of pork, and I was intent on putting my money where my mouth was, which was usually on some form of encased meat. As I've discussed, all "game animal" hunting in the United States is highly regulated in order to ensure the stable population numbers and health of those animals–in contrast, feral pigs have no bag limit, meaning you can kill as many as you want, and no season, so you can hunt them year-round. They're also inexpensive to hunt. A deer tag in my home state of New York would cost me $22 as a part of my regular hunting license. Someone from out of state would pay $100 for the same license. In Texas, Chris doesn't have to pay a dime to hunt pigs (outside of his regular license), and I only had to pay $45 as an out-of-state hunter. They're basically an unlimited food source–you could never hunt or even trap enough of them to dent their population. And in Texas, they try. To truly attempt extermination, they'd need to rely on massive round-the-clock culls by professional trappers and sharp shooters in helicopters. The same goes for most invasive species–there's lots of little things we can do to help, but proper extermination requires concentrated funding and dedication from professionals.*

Hunting or trapping invasive species is one way to put some food on the table and help out local ecosystems at the same time. You can hunt feral pigs in at least thirty-five states, soon to be more, as the US Department of Agriculture considers their population to be "between six and nine million and rapidly expanding." Jackson Landers, who wrote about hunting deer for food in 2011, wrote a follow up in 2012 about hunting invasive species, called *EatingAliens*. In it, he travels the country detailing his hunts for some of the most notorious invasive species in America. If one were so inclined, they could hunt and fish for nothing but invasive species while helping the environment twice over with those eating habits. You could travel to Florida to hunt python and iguana, spear lionfish all along the southern East Coast as far north as North Carolina, and fish for Asian carp across most of America. Hawaii, the United Kingdom, New Zealand, and Australia are all popular hunt destinations that love when tourists pay to help with invasive animal management. Japan is having serious hog problems too and attempting to boost hunter numbers on the island of Hokkaido.

---

*An *introduced* species is one that's been brought to a non-native habitat accidentally or on purpose. Many began spreading around the world with long-ago colonizers bringing domesticated or wild animals to be farmed or hunted. Today the pet trade is a main originator of introduced species. Others hitched a ride in a boat bilge, transom well, or maybe a stack of bananas. An *invasive* species is an introduced species that has posed a serious threat to local native wildlife and habitat since its introduction. All invasive species are introduced, but not all introduced species are invasive.

In New Zealand, the introduced Himalayan tahr thrives in the mountains where it has no competition or predators, and it doesn't negatively impact the ecosystem there. Conversely, the common brushtail possum, which was brought to New Zealand from Australia, wreaks havoc on native small animal populations and forest canopy ecology—it is invasive. Iguanas and pythons are dangerously destructive invasives in southern Florida—they've cleared the Everglades of 90 percent of its native mammal population (pythons alone are responsible for the disappearance of 99 percent of raccoons, 98 percent of opossums, and 87 percent of bobcats; marsh rabbits, cottontails, and foxes have completely disappeared). Equally non-native chameleons introduced in Florida pose no known threat to the area's local native species or habitats, hence no campaign to remove them (Florida has to deal with about 180 other introduced species of reptiles and amphibians first).

In some states, there are bounties on invasive species, so you're the one getting paid. This helps create incentive for hunters and trappers to target species that normally aren't on their radar. People like to eat pork, so you don't need to (and in some instances shouldn't) incentivize hog hunting further. But critters like swamp rats have less value to hunters, so a little monetary push helps make them more appealing. The bounties range from five bucks for a Louisiana nutria to a full-time job hunting pythons in Florida. Python hunters can get up to $15 an hour, with bonuses from $50 to $150 depending on the snake's size, and a further $200 for an active nest. There are only fifty positions available for the full-time contract, but recreational hunters can still hunt pythons without needing a permit.

The amount of money paid out to those recreational hunters and trappers is less than the cost to taxpayers would be if Fish and Wildlife departments had to do the same job on their own, so it's worth it for the government to outsource the work. There are bounties on fish too. The beautiful but ecologically nightmarish lionfish was introduced to East Coast waters via released pets and, because they don't have any natural predators in the locations they invaded, spread quickly, eating and eating and being eaten by nothing. Fortunately, they're slow moving and easy to spot, so they've become a popular target for both local and tourist spear fishers. Today, they're subject to the occasional bounty, and coastal communities are trying to make them more of a restaurant staple.

In the Pacific Northwest, the pikeminnow is the number one target of control efforts. They aren't invasive, but dams and other hydraulic systems have changed their habitat, making it easier for them to prey on baby salmon (smolts). The goal of the Pacific Northwest program is not to eradicate pikeminnow but to reduce the numbers of large individuals so more juvenile salmon can make it out to sea. The more you catch, the higher the bounty. As of 2021, 1–25 fish gets you $5 per fish, 26–200 goes up to $6 per fish, and 201 or more climbs to $8 per fish. Catching a fish with a tag on it gets you a $500 bonus. The program has been working. The Washington Department of Fish and Wildlife reports they've seen a "40 percent drop in predation on juvenile salmonids" since its inception. All those bounty dollars paid to anglers still accounts for less money than the taxpayers would ultimately have to shell out if they were suddenly the ones on the hook to protect Washington state's culturally, economically, and environmentally significant salmon population.

Wildlife departments have to monitor all bounty programs carefully to make sure animals aren't being captive raised or otherwise encouraged to breed just for the bounty on them. The danger of creating a market for invasives that might encourage profit-seekers to increase a population's numbers is real but less of a concern than it has been in the past, partially due to the incorporation of the tourist economy in the removal of invasives. People who aren't part of a local or traditional hunting or angling community, who may not know all the history or economic benefits of that lifestyle, can still jump into a lionfish spear hunt on their vacation and feel good about their contribution to the regional ecosystems they're visiting. Guides who profit from the tourist economy might benefit from the pigs or the iguana but not enough to incentivize adding to the species (especially since the pigs and iguana do an adequate job of reproducing on their own).

Hunting invasives doesn't produce as much revenue for Fish and Wildlife services or their conservation projects as hunting traditional game. Because the idea is to get rid of the invasives, you don't want to make barriers, like charging too much (or anything at all), for doing what amounts to a favor for local ecologies. But even though Fish and Wildlife isn't making money, they are saving money, enabling them to focus their funding and the time of their trained professionals on big-picture stuff, while the public helps out with the grunt work.

Hunters and trappers are rarely able to do everything needed to manage invasive species on their own—there's just too many of them, no matter how many pig dinners you have (which should tell you something about how bad it could be if hunters and trappers weren't controlling them at all). But increased literacy about the topic could draw more nontraditional hunters to the dinner table. And even though it's unlikely we'll eat our way out of invasive animal–related problems, invasives do provide a gateway hunt and an introduction to understanding wildlife management. Invasive species could be the ones that get people outdoors, reacquainted with ecology, and interested in providing for themselves in a way that's perhaps less complicated to feel good about than a deer that looks like Bambi. That's what feral hogs were for me.

In the lead up to my Texas trip, the process of taking Hunter's Ed, researching hogs, and packing for my trip still felt far removed from my final goal of actually killing an animal for meat. I love the outdoors, travel, food, and a day at the shooting range, and so far, that's all this trip had been. Hang out with a friend, get some good barbecue, do a little target shooting. I knew in the back of my mind I had come to Texas to kill an animal for the first time, and at some point, in the next few days, that might happen. I tried not to think about it. I figured if I got to a point where I had my rifle trained on a pig and I just couldn't pull the trigger, then I wouldn't. But the hypocrisy of having that thought while my mouth was full of deeply satisfying smoked pork ribs was heavy enough for me to think I could.

The next day we loaded up the truck for the four-hour drive north from Houston to the land we'd be hunting. We drove around Chris's portion of the property checking camera traps and feeders. All the feeders we checked were either broken or empty, and we spent the afternoon fixing them. We looked at the neighboring land that belonged to Chris's dad, but the feeder there was empty too, so Chris decided any hunting would have to be put off till the following night after we'd had a chance to visit the grain lot and refill them all.* That meant we lost two hunting opportunities out of four: that night and the following morning. Which was a blow to my odds of

---

*Tanks of dried corn and mixed grains with a timed release are just one form of baiting—depending on where you hunt and what you're hunting for, bait could be as simple as a salt lick left in the woods for deer or a pile of doughnuts and fish guts (and all kinds of stuff) for black bears. Hunting over bait is legal in some form in almost half the country, but laws vary greatly depending on the state and species. In this case, I found myself with mixed feelings. Because I was hunting wild pigs, I wasn't particularly concerned about the fairness of the fight—feral pig populations need to be kept in check, and if having a pile of food in a strategic location led to a pig giving me a more clean and open shot, then I figured I should be thankful for that. The pigs could have still seen us going to the blind or smelled us at any time. So how much of a "cheat" was it if the bait was there and the pigs weren't?

Rick Taylor, a wildlife biologist who wrote the book on feral hogs for the Texas Parks and Wildlife department, says "There is currently an estimated population in excess of 1.5 million feral hogs in Texas." So, bait or not, I'd have been happy to take a couple off the landscape. In the end, my personal opinions on the matter broke down along invasive versus native species lines. I wasn't interested in hunting native species over bait. But maybe I'd change my tune the more my diet depended on wild game?

seeing or shooting a pig. I was encouraged by the camera trap photos though. I couldn't believe how many pigs there were–they barely all fit in the frame.

We got into the blind by 5 p.m. the next day and sat there silently until about 8:40. Since this was private land, Chris had a permanent blind set up. It was a little box with a door, tall enough to stand up in, resting on two trailer wheels, and with a long, slim horizontal door on each side to slip the muzzle of a rifle out. Inside there were two beat up old office chairs, one of which squeaked relentlessly if you moved so much as a millimeter. It was hot too. Opening the back door rifle slot would have pulled a nice breeze through, but you can't take the chance that you'll blow your scent around or create shadows if you move. The hogs didn't show up, but I did get to watch two raccoons climb over one of the feeders further away from us. Three bucks passed through the area too, young guys probably only one and a half years old. It was definitely fun to spy on animals who didn't know I was there.

The next morning would be my last opportunity before we had to drive back to Houston, so we got up early to get into the blind before sunrise. I felt some pressure, which I had no clue what to do with considering I couldn't make the pigs show up by force of will. I'm a night owl, and, left to my own devices, I'll go to bed after midnight, so I'm not often awake to see many sunrises. But I'm quick to admit (based on the increasing amount I've seen since taking up hunting) that they are vastly superior to sunsets. I remember the quality of the light changing from black to silver to blue, before any of the oranges or yellows appeared. As soon as the sun was high enough to warm the ground, thick billows of steam started coming up from the earth and slithering into the open window of the blind. It looked like the kind of smoke that would emit from the nostrils of a sleeping cartoon dragon. It was simultaneously meditative and exciting.

I watched everything–the light, the steam–it felt like I looked at every blade of grass through my binoculars. I must have had first-timer fever. I was surprised at how I'd never felt my heart beat faster than when I was doing nothing more than sitting still and just waiting for something out of my control to maybe happen. My adrenaline pumped for four hours straight, and all I was doing was looking at a sunrise and watching the same two raccoons stick their little arms up the feeder chute to feel for crumbs. I still don't know if all my nerves were a product of wanting to kill a boar or from *not* wanting to. And even with all the hunts I've been on since, hunting has yet to lose its jitter-inducing qualities (though they come in more predictable waves and spikes now). At some point, one of the young bucks walked closer to the blind, and, getting a whiff of us, stopped dead in his tracks and looked right at us for what felt like forty-five minutes but was probably closer to four. I have never put so much energy into not moving. As Chris kept his head down, buried in his phone, under the maxim that a watched pot never boils, I bet I burned the same number of calories as an Olympic swimmer just sitting there staring back at the buck.

I've talked about the importance of funding to conservation, and it can't be overstated, but to be worth anything, all that money needs to funnel into a system that actually works. The North American Model of Conservation is a set of seven guiding principles that was shaped by and has shaped nearly every wildlife-related law in America, and some international laws as well. There's no single person responsible for their creation. The tenets of the model morphed over time from those early and various wildlife laws that began to be enacted in the late 1800s. However, conservationist, author, professor, and awesomely named wildlife biologist Valerius Geist (who passed away in 2021) is credited with formally phrasing and organizing "The Model" in 1995. He and conservation educator Shane Mahoney edited a book about the model with eleven other authors in 2019. Currently, in the language of the US Fish and Wildlife Department, the guiding principles are as follows:

1. Wildlife is a Public Resource
2. Markets for Game are Eliminated
3. Allocation of Wildlife by Law
4. Wildlife Can Only be Killed for a Legitimate Purpose
5. Wildlife Species are Considered an International Resource
6. Science is the Proper Tool for the Discharge of Wildlife Policy
7. The Democracy of Hunting

1. Wildlife is a Public Resource

In the United States, wildlife is considered a public resource, independent of the land or water where wildlife may live. Governments at various levels have a role in managing that resource on behalf of all citizens and to ensure the long-term sustainability of wildlife populations.

Nobody owns wild animals. By making the animals themselves the resource, we ensure their protection no matter where they're walking, flying, or swimming. So even if a group of mule deer are hanging out on somebody's private land, that doesn't mean the landowner has any rights to them. The animals are held in trust by the government so they can be managed and protected for the long-term benefit of everyone, including the animals themselves. This principal is often referred to as the Public Trust Doctrine. The origin of this one is credited to an 1842 Supreme Court ruling where a guy wanted exclusive rights to the oysters living on the mudflats of his New Jersey property. The court said, no, the oysters belonged to the state. The oysters were for everyone.

2. Markets for Wild Game Are Eliminated

Before wildlife protection laws were enacted, commercial operations decimated populations of many species. Making it illegal to buy and sell meat and parts of game and nongame species removed a huge threat to the survival of those species. A market in furbearers continues as a highly regulated activity, often to manage invasive wildlife.*

**Decimate*. As you may remember from high school Latin or Roman history (or your most pedantic friend or colleague), the prefix *deci* means "tenth," and the original definition of *decimate* meant to kill one out of every ten Roman soldiers from a group of around 400–500. I can neither confirm nor deny that the passenger pigeon was a member of the Roman army, but I can say conclusively that more than one out of every ten was killed. However, passenger pigeons do fall under today's common and acceptable definition of decimate, meaning to destroy a great number of. So do bison.

As far as the seven tenets go, this is one of the more interesting, because it has so many specific exceptions. A basic modern example would be "game meat" at a restaurant. Based on this rule and most state laws that follow it, you cannot serve or sell wild animal meat at a restaurant or business. That would mean you're making money off Mother Nature with no system in place to get the profits back to her. When you see venison, elk, or alligator on a menu, that's a captive, farm-raised animal just like a cow or pig. It may be "exotic" because you don't see it on menus often, or called "game" because it might be a game species, but it was never wild.* There are some exceptions for invasive species–brands like Creminelli make a salami from feral Texas hogs and Force of Nature has an entire line of Texas wild boar products (they even sell t-shirts that say "Eat More Invasive Species").

3. Allocation of Wildlife by Law

Wildlife is a public resource managed by the government. As a result, access to wildlife for hunting is through legal mechanisms such as set hunting seasons, bag limits, license requirements, etc.

The public has legal access to wildlife. But it's up to the government to organize what that access looks like. The goal is the long-term health and stability of wildlife populations, so that means hunting laws that change not only state by state but also region by region, based on the most current applicable local data. Hunting seasons vary place to place because they must be set in accordance with the life cycle of those species to ensure they can continue to reproduce year after year.

---

*In Britain, you *can* sell game meat at a grocery store, and those animals are wild, not farm raised. Game bird chicks (like pheasant and partridge) are often raised briefly before they're released into the wild to aid in a higher rate of survival and ensure sustainable numbers of birds while keeping human intervention to a minimum. But venison from the various deer species in Britain are totally wild. Mark Staples, the commercial director of the British Game Assurance, told me he's seen an uptick in game meat sales. A combination of people seeking out more environmentally sustainable meat sources, as well as people using the extra time at home during the pandemic to experiment with game meat cooking (which is leaner than commercially raised farm animal meat and therefore requires some cooking adjustments).

Bag limits, like hunting seasons, are also subject to change. Turkey population not looking so robust this year? Drop that bag limit from two to one. Whitetail deer so numerous the local highway is a deer graveyard? Extend the length of the season or bump up the bag limit. State Fish and Wildlife biologists use the data they've spent the whole year collecting to make sure seasons and bag limits keep wildlife populations stable or growing. They're responsible for adjusting them with that in mind.

License requirements are generally less impactful than bag limits and don't have a reason to change annually. Some states, after seeing the success other states have had, may wish to adopt an apprentice license if they don't already have one. In 2021, New York state dropped the deer-hunting age from fourteen to twelve, if accompanied by an older licensed hunter. This may not seem like a big deal, but two years is a world of difference for kids who have grown up in a hunting family. The sooner you can give kids that level of responsibility, the sooner passionate outdoor appreciation becomes personal to them.

4. Wildlife Can Only be Killed for a Legitimate Purpose

Wildlife is a shared resource that must not be wasted. The law prohibits killing wildlife for frivolous reasons.

Eat what you kill. There are specific laws that dictate how you must use the animal you're hunting if you're successful. These are called wanton waste laws. Most states stipulate that you must take a percentage of meat, but they articulate how in a variety of ways. Alaska is quite detailed–all useable meat must be removed, including from the neck and between the ribs. Montana says all edible parts of a bird must be used–you can't just take breast meat. A "legitimate purpose" can also be hides and pelts or population control. Basically, you can't just go around shooting things for no reason. Each state has their own legal way of saying "don't be an asshole."

5. Wildlife Species are Considered an International Resource

Some species, such as migratory birds, cross national boundaries. Treaties such as the Migratory Bird Treaty and CITES [the Convention on International Trade in Endangered Species] recognize a shared responsibility to manage these species across national boundaries.

The Migratory Bird Act was put in place a little too late to help out the passenger pigeon—they went extinct in 1914, and the population began collapsing long before that. The MBA wasn't established till 1918. Now, Canada, the United States, and Mexico (and sometimes Russia and Japan) are bird buddies, and we all work together to keep tabs on migrating bird populations and their health (along with some other critters that may cross international boarders). Our partnerships and commitment to information sharing represent crucial data sets when determining the most up-to-date hunting and wildlife protection regulations. These international partnerships aid climate change research as well. Watching the international migration patterns of birds shift as they adapt to a warmer world has broader implications beyond hunting seasons. The US Fish and Wildlife Service estimates "that more than 300 species will be driven to smaller spaces or forced to find new places to live, feed, and breed over the next sixty-five years."

My favorite part of the MBA is the original legal cartwheels undertaken to ensure the protection of the migrating birds under a variety of circumstances. The protections include the establishment of a federal prohibition, unless permitted by regulations, to "pursue, hunt, take, capture, kill, attempt to take, capture or kill, possess, offer for sale, sell, offer to purchase, purchase, deliver for shipment, ship, cause to be shipped, deliver for transportation, transport, cause to be transported, carry, or cause to be carried by any means whatever, receive for shipment, transportation or carriage, or export, at any time, or in any manner, any migratory bird, included in terms of this Convention... for the protection of migratory birds... or any part, nest, or egg of any such bird." I like to picture a sweaty, frenzied lawyer banging on a typewriter, smoking the gnarled stub of a cigar, and muttering, "Let's see them try to get around this!"

The Convention on International Trade in Endangered Species (CITES) got its start in 1963 but wasn't fully hashed out until ten years later. Currently, around five thousand animal species and twenty-five thousand plant species are protected by the treaty. It's a huge undertaking because wildlife crime (illegal trade) is so widespread and profitable. On its website, CITES explains, "Annually, international wildlife trade is estimated to be worth billions of dollars and to include hundreds of millions of plant and animal specimens. The trade is diverse, ranging from live wild animals and plants to a vast array of wildlife products derived from them, including food products, exotic leather goods, wooden musical instruments, timber, tourist curios and medicines. Levels of exploitation of some animal and plant species are high, and the trade in them, together with other factors such as habitat loss, is capable of heavily depleting their populations and even bringing some species to extinction. Many wildlife species in trade are not endangered, but the existence of an agreement to ensure the sustainability of the trade is important in order to safeguard these resources for the future."

White rhinos are endangered, but CITES allocates hunting permits for mature bulls past their breeding prime, because the money paid by big game hunters goes toward protecting the species from poachers. The organization Save the Rhino explains the importance of the legal trade using Namibia's success with rhino breeding as a model, "High costs can reduce private ownership, which has helped to increase Namibia's rhino population in the first place (nearly 80 percent of Namibia's white rhinos are privately owned). With high-security costs, the challenges that come hand-in-hand with rhino poaching, and the intense, ongoing drought, private rhino owners need an incentive to keep rhinos."

The Endangered Species Act has done wonders for bringing some species back from the brink. According to US Fish and Wildlife, 99 percent of all species put on the list haven't gone extinct. But we can't forget that, if all goes well, *endangered* is meant to be a temporary status. Biologists are (or should be) in charge of deciding when a species has recovered based on population numbers, distribution, and its current projected range, as well as an area's carrying capacity. Once a species has recovered, then biologists can resume normal management (which can include hunting) so that species can continue to thrive and hopefully not have to go back on the list.

Historical ranges of species' populations and distribution would be fun guides to follow when determining recovery, but that's not always possible as people dominate more of the landscape. There was a push from conservationists at the Rocky Mountain Elk Foundation to reintroduce elk to their historic range in the vast Adirondack mountains of New York, but between homeowners concerned about their gardens getting nibbled and auto insurance lobbyists worried they'd have to pay damages from more animal strikes, it never happened. (Pennsylvania, Kentucky, and Virginia* all got their elk back, so I'm kinda bitter.)

6. Science is the Proper Tool for the Discharge of Wildlife Policy

> *In order to manage wildlife as a shared resource fairly, objectively, and knowledgeably, decisions must be based on sound science such as annual waterfowl population surveys and the work of professional wildlife biologists.*

Be still my heart. Wouldn't it be nice if all lawmaking was held to this rather common-sense standard? This is one of the reasons wildlife management in America has been so successful—we do what the scientists say. It's the root of wildlife and wildland management in this country, but I can understand how that might be confusing. Why do we need to "manage" anything? Can't we just leave nature alone for once and let it do its thing? Bizarrely, the answer is only "sort of." We all, human animals and other animals, live on the same planet, but humans keep turning shared habitat into anything and everything but. We can only shrink

---

*Elk were reintroduced to Virginia in 2011, and by 2022, there was enough of a stable population to hold the state's inaugural elk hunting tag lottery. Virginia opened the lotto to both in-state and out-of-state hunters hoping to score one of the five allotted tags, raising over half a million dollars for the Virginia Department of Wildlife Resources.

wild habitats so much before they require messing with. The earliest representatives of our modern species, *Homo sapiens*, appeared on the scene around 315 thousand years ago, and we have spread into all the nooks and crannies of the earth and disrupted ecosystems to such a degree that now we must continue to meddle with some places if we want them to maintain the level of harmony we associate with a wild place today. There is no part of Earth that hasn't been altered by humans already.

The Smithsonian Institute reported that humans caused a "major shift in earth's ecosystems" around six thousand years ago, and the planet was forever transformed by human activity around three thousand years ago. Now, the places many Americans think of as the most "wild," like National Parks, are often the most groomed outdoor spaces. As Emma Marris puts it, "National Parks are heavily managed. The wildlife is kept to a certain population size and structure, fires are suppressed, fires are started. Non-native species are removed, native species are reintroduced. It takes a lot of work to make these places look untouched." It's basically an outdoor zoo. There's nothing wrong with that except for how it informs the way we interact with nature. Some areas are managed with the intention of "just look," and some with the intention of interaction. Both are valuable. But interaction often brings more connection, and connection brings greater advocacy.

Some seasonal behaviors of animals are thought to be changing due to the climate emergency, and wildlife management is adapting. The hunting of Canada geese and other migratory birds might look different in the future if warmer temperatures and food availability continue to alter how long they decide to stay in one place during the migration period. Changes in light, temperature, and food supply are all triggers that tell the geese when to migrate, but we're already seeing different patterns in migratory bird behavior as temperatures and food supplies shift with climate trends. A study from the University of Exeter found that while the warmer temperatures were leading female Atlantic Brant geese (called light-bellied Brent geese in the United Kingdom) to produce more young, it was also putting the females at greater risk of death. The females would use up valuable strength and energy to produce more eggs than usual but then they'd become extra vulnerable to predation while nesting. And while having more little Brants running around sounds great, they won't be able to hatch without mom keeping them warm, so that's a breeding female and her whole brood that's in danger of being wiped out.

Fire management is another good example of how cycles of human meddling were, and have become, necessary. Native Americans used small, prescribed fires to regenerate the soil in certain areas, which led to new growth of forest understory and protection from larger fires. But the practice was outlawed by a US government that thought using fire was primitive and dangerous. Thus began some one hundred years of fire suppression strategies in America that have turned much of the country into a giant tinder box. Park, forest, and conservation agencies started to change their tune in the late 1960s, but now, with people's homes everywhere and landscapes becoming hotter and dryer, prescribed burns have become more frightening. It's harder to manage a fire in such conditions, which is why the vocabulary has shifted from *controlled burn* to *prescribed*. Fire is hard to control these days.

The most unfortunate aspect of Rule Six is how it's become the most endangered. Increasingly, public opinion is nosing its way into wildlife management. What's surprising is that it's well-meaning animal advocates doing the nosing–people who might broadly categorize themselves as environmentalists–actively seeking to remove decision-making power from professional, pro-science, wildlife and environmental specialists. I've sat in on a number of public hearings on topics like bear hunting in California, and public opinion tends toward the poorly supported and emotionally driven. Keeping science in the hands of scientists shouldn't be controversial.

7. The Democracy of Hunting

In keeping with democratic principles, the government allocates access to wildlife without regard for wealth, prestige, or land ownership.

Still sore from the days of kings owning all the hunting land as well as the wild animals themselves, policy makers in the early days of the United States made a concerted effort to ensure that hunting and fishing were available to everyone who wanted to partake, regardless of how rich or connected they were or how much land they owned.

Access is a buzzword in the hunting community, but it doesn't have to be hunting specific. It refers to having access to wild land (important to hunters because if you don't have access to the land, then you don't have access to the animals). Some public lands are "landlocked." Meaning those public lands are completely surrounded by private land, making them totally inaccessible to . . . the public. A digital mapping and navigation technology company called onX partnered with the Theodore Roosevelt Conservation Partnership from 2018 to 2020 to produce a series of reports aimed at determining how much public state and federal land is inaccessible to the public across the country. California for example has 592 thousand acres of landlocked public land, New Mexico has 1.83 million acres, Nevada has over 2 million, Montana has over 3 million, and Wyoming has over 4 million. The total for the twenty-two states they've profiled so far comes to 16.75 million acres of landlocked public land. The TRCP continues to work on unlocking public lands and modernizing public land data so more people can easily find places to get outside.

In the hunting space specifically, the "privatization of wildlife access" is spreading more rapidly and spreading controversy along the way, with private landowners selling hunting and tag rights on their property for a profit. It's a double-edged sword–if a landowner can sell hunting rights, then they are incentivized to maintain habitat for wild species. That's better than selling drilling or mining rights, but it does favor those who can pay for access. There are also programs that focus on public access to private lands, as well as habitat incentives like the Farm Bill that aim to support private landowners and help the hunting and non-hunting public gain access at the same time.

It should be noted that many a private landowner is happy to have hunters come hunt their property for free as long as they ask nicely, but incentives from one place or another are an important way to ensure wild lands maintain value. Hunting guide and former forester and wolf biologist Adam Gall hunts public and private lands and summarized what he's seen in the field saying, "The issue of paying for access to private land is complicated, but there's lot of upside to the concept, as it creates incentives for private landowners to maintain wildlife and vital habitat, especially critical winter range. And I think that's worth mentioning. Many folks who might be 'anti-ranching' or 'anti-cowboy' simply don't understand that if those folks go out of business, that land is almost a given to get subdivided. Kiss your wildlife goodbye."

Chris is a member of a property lease with about ten other people. Despite how large and hunter friendly the state is, Texas actually ranks number forty-seven in terms of available public land, so if you want to hunt regularly, you better be a part of a lease or know someone who is. The property is huge–the landowner has a cattle ranch on one portion of it, and each member gets over 100 acres to hunt and manage themselves. This is a pretty typical arrangement in Texas, though, of course, property sizes vary. Managing your portion of the land can mean clearing out old brush, planting trees and shrubs that the deer and turkey like best, or maintaining feeders. Public land advocates are probably finding all the hairs on their necks standing up, but because Texans have so little public land available to them, leases have become the norm.

I didn't shoot anything on my first hunt. I didn't even *see* a hog while I was out there. But that's not unusual. The National Deer Association (formerly Quality Deer Management Association) is a science and education group dedicated to perfecting whitetail management in the United States, and in 2017, they put the average deer hunter success rate at 41 percent for a hunter's first deer of the season. Wild pigs have an even better sense of smell than deer and can be even more skittish. I have no doubt the uncertainty of success is part of the long-term appeal of (addiction to?) hunting, or at least what keeps hunting challenging and exciting no matter how many days you spend doing it. It's what makes hunters smile in a photograph with their quarry. Hunting is not easy, and success is not a given, which is why it's so rewarding when a hunter does bring home the bacon. What if every time you went to the market, there was only a 41 percent chance you'd be able to buy meat? I bet you'd appreciate it more. I bet it would become special.

Imagine the addictive quality of gambling or the feedback loop that keeps you checking your phone every ten minutes–unlike your mathematical chances of winning at the casino or algorithm-driven interactions on social media, you can get better at hunting. Better, but success is never guaranteed. If you do get "lucky," that's meat in the freezer, and whether you're successful or not, you've still paid into the conservation funding system and spent time outside. These are notable consolation prizes and the reason so many hunters take issue with the term "success." It's always a good day outside, even if you don't connect with an animal. I can't say as much for losing money at the casino.

I couldn't tell how I felt about being in a blind for this hunt. In general, I accepted that it was the most practical and common method of secreting oneself for this type of hunting. My goal was to kill a pig and be responsible for turning that pig into pork. I was happy to not have to think about stalking or the myriad other things that go hand in hand with that style of hunting. Not yet at least. It reminded me of being a river guide when the topic of outdoor survival would come up. This was around the time survivalist TV shows like *Survivorman* and *Man vs. Wild* were gaining popularity. Something about being outdoors with people always brings out romanticized wilderness survival stories, and people in the boat would inevitably start talking about those shows. I think that type of TV is fun but can also do a disservice to what should be the safety concerns of regular people spending time outside. You watch these shows and suddenly you think survival is an extreme, reserved only for SEALs and doomsday preppers, or that it means being thrown out of a helicopter onto the arctic tundra with nothing but a roll of dental floss and a ballpoint pen. In reality, survival is just the more badass name for wilderness safety and preparedness–filtering water, knowing you need to warm up before you get cold, knowing how to start a fire, and having the tools on you to do it. But that's all way less fun to watch on TV than whatever Bear Grylls is gonna do with that floss. Shows like *Alone* have done a better job of showcasing survival skills in less dramatic circumstances.

I was fully aware of this mental trap but couldn't help expecting or wanting my first hunt to be filled with the type of adventure or danger I had read about as a kid or seen in the movies or on TV. I wanted to try and sneak up on the hogs and have one come out of the bushes like the raptor in Jurassic Park, alerting me that it was time to fight off an attacking swine horde.* What that really boiled down to was what I perceived at the time to be a fairer fight. Maybe I wouldn't feel so bad about killing an animal if it was trying to kill me first. And no hog was going to break into my elevated, metal hunting blind.

At the end of the day, it didn't matter–no hogs ever showed up. Maybe we got to the blind too late, and they'd already moved on. Or maybe they were aware of us as we arrived–many hogs, especially those in Texas, have become nocturnal as a response to hunting pressure (in turn, Texas has opened hog hunting hours to any time of day or night, a stark contrast to most game animals that have strict daylight shooting hours within their seasons). Maybe the pigs smelled us or were put off by all the commotion we had caused the last couple days around the feeders. Maybe they were just hanging out elsewhere. The point is, you don't always get one, even if they're used to a blind being there or know there's food when they hear the sound of a feeder go off. It's not a sure thing. I realized the fight *was* fair, because even though I was hiding in a blind and had a rifle, I would never have the hearing, sense of smell, or instincts of my animal quarry.

I was both relieved and disappointed at the outcome of that hunt. Relieved I never even had the opportunity to pull the trigger, but disappointed in how that led to a fallout of missed lessons and emotions. Even if the pigs had shown up, there was no guarantee I'd have found a comfortable shooting opportunity. So, one lost lesson was choosing my quarry from the group, knowing his range, training my rifle on him, and picking my shot placement. After that, I lost the opportunity to know what it felt like to pull the trigger on an animal and how I'd feel after. Then the practical basics of gutting and processing my kill. And the worst lost lesson–what would it feel like to have thirty pounds of wild meat in the freezer that I was finally responsible for? My empty freezer was as clear of an answer as I needed, and I started planning my next hunt.

---

*PSA: Do not fight a wild hog in hand to hoof combat. Without canine assistance, you're unlikely to win.

6

Hunting Solo

Bows, Guns, and Politics, Oh My

I met my hunting buddy Mike at a small hunting and fishing event in New York City. We hit it off right away (probably because we both speak in movie quotations like fourteen-year-old boys). A couple years later, he and his wife, Colleen, invited me to their place in the Catskills for Thanksgiving and the fall archery and rifle season. I could not have asked for a better gift from anyone. I spent five weeks living in their downstairs guest apartment, and in the mornings Mike and I would go hunt his archery spot together until I finally got my own bow. Then he was generous enough to find me a little site I could hunt myself.

I would wake up earlier than I had to (an unheard-of occurrence) and try to sneak out of the house without upsetting the dogs or waking their toddler (perhaps the most hair-raising part of the day). Then I'd walk down the road in pitch darkness with my bow and backpack to a small bridge, just big enough for one car. I stood in the middle of the bridge to make all my final adjustments before walking to my hunting spot, using the sound of the rushing water under me to mask the noise of fidgeting with my bow or dripping some cover scent on my boots. I loved settling into my spot so early. I let myself zone out and daydream in the darkness, waiting for the first hints of light to show up behind the mountains. Groups of does would filter through the area and send me on extended adrenaline highs, while I sat perfectly still waiting to see if they'd make their way close enough to me.

In the area I was hunting, does are legal during that early part of the season, before the rut (breeding period). The population can sustain and benefit from losing some each year, but once breeding begins, does go off the market. I can't remember how many mornings and afternoons I sat there, but even though I never had a deer come in close enough to shoot with my bow, I don't regret a moment out there and I was never bored. I loved watching the woods, and, because I didn't have to focus on my footing or the next trail marker, I was able to take them in better than any time I'd gone hiking. Once, a chipmunk who either didn't know or didn't care that I was there, ran back and forth over the tops of my boots while I rested them on a log.*

A hunting season is created in accordance with a species's breeding cycle. Once the season is set, it's broken up into periods that take hunters' tools into account. Bows, of any kind, are harder to use than rifles. They require you get closer to your target. That's why hunting seasons throughout the country start with archery. The advantage is for the prey, let's say it's deer, and also the hunter. The deer haven't been hunted in the last six to seven months so they're slightly less wary, and the archer might be able to get closer. Once rifle season starts, the deer have become more careful,

---

*Another time, a plump ruffed grouse flew in behind me and pecked around the tree I was sitting against before flying up onto a branch six feet in front of me. I was giddy. I had wanted to see a ruffed grouse the moment I moved back to the Northeast. They're known as the King of Gamebirds to hunters for a few reasons: 1) the males have a prominent ring of feathers surrounding their neck that they'll puff up and extend when attracting a mate or defending territory—the name ruff and the nickname king comes from the ruffed collars worn by European aristocracy of the late sixteenth century; 2) ruffed grouse are true forest birds, unlike most popular upland game birds that live on open plains and in grain fields; and 3) the males make a drumming sound which sends chills up the spines of forest birders and nerdy hunters the way a lion's roar might in Africa. The sound is achieved by the bird bracing its tail down and beating its wings so hard the resulting vacuum creates mini sonic booms! I wrote in my journal how excited I was to see one up close, like a teenager in the grips of Beatlemania.

but the rifle hunter can shoot from a farther distance. This keeps the deer's advantage in line with the hunter's.

If you really delve into the culture, weapon choice is where you'll find a lot of squabbling among hunters. Bow versus rifle and bow versus bow. There are plenty of traditional bows, compound bows, and crossbows out there to suit every archer, but if you plan to only use a bow, your hunting seasons will be limited to the dates bow hunting is allowed, as will the type of game you can hunt. Most small game hunting benefits from a shotgun or small caliber rifle, but you can hunt rabbits and squirrels with a traditional bow if you practice. If you'll recall, I used a crossbow to get my first deer. Vertical bows, like traditional and compound bows, take an immense amount of practice to become proficient with, and people can still easily make bad shots using them, even with years of practice under their belts. Many new hunters choose to pick up crossbows because of the reliable accuracy they provide and because they're physically easier to use than a vertical compound bow, but they're becoming popular with hunters of every stripe now. Many hunters use crossbows when they start to get bad shoulders and can't pull back their compound bows anymore. They're also easier for kids and folks in wheelchairs or with different physical abilities to use, which means more people can enjoy the archery season.

A traditional bow could be a longbow or a recurve. Both have a single string attached to the far end of each limb. Pull back the string, create tension, release the string, and the stored energy will propel your arrow forward. The farther back you pull, the more energy you'll put into your arrow. Traditional bows can be made from wood like they have been for thousands of years or out of schmancy synthetic materials. You can absolutely hunt with them, but they're more often used for target shooting. Compound bows are more popular to hunt with. They have extra strings and cables connected to cams at the end of each limb. This block-and-tackle-style pully system allows the bow to store more energy in a shorter, more compact package and to "let off" weight at full draw. With a traditional bow made for sixty pounds of draw weight, when you draw the string back, you're holding all sixty pounds till you release it. If your compound bow is set to sixty pounds of draw weight, but it has a 75 percent let off, then you're only holding fifteen pounds when you're pulled back at full draw.

Hunting with a vertical bow is hard (which is why people love it)—having to get closer is one part of it, but it also takes a crazy amount of practice to shoot well. Once you've drawn your bow back, you don't have that much time before you have to release the arrow. The bow (especially a traditional one) is under tremendous tension, and it can be difficult to hold steady because of that. Crossbows, on the other hand, will hold the string under tension for you for the entire duration of your hunt. I love learning a craft and trying to perfect my skill at something, and I'm definitely in the market for a compound bow, but in the meantime—until I feel like an Olympic-level eagle eye—I'm inclined to opt for a more consistently accurate weapon I feel confident using.

Caption TK

Crossbows are a hot topic in the hunting world because some vertical bow users (traditional and compound) don't want to share the archery season with a bow that's more user friendly than theirs. They're basically worried that, because shooting a crossbow is easier than a compound bow, the bow season will become flooded with yahoos who used to only hunt the rifle season. I don't blame them for the concern, but the crossbow has proven to be a powerful tool to get new hunters into the field, especially for those who are gun-shy, and for a lifestyle begging for more representatives, I think that's worth consideration. Not to mention crossbow users often become vertical bow users. It's a gateway bow. Personally, I have no problem splitting the season. If vertical bow users want first crack, that's fine by me, I'll wait till my compound arrow groups are up to snuff before I join them.

Caption TK

So. Guns.

[Stares at blinking cursor. Sighs. Talks to self.]

You could be writing fan fiction about Quint and Hooper right now, but nooooo, you decided to discuss guns, shooting animals, and politics.

[Rubs temples.]

I'm a pretty progressive dude. I also enjoy owning and shooting guns. I do not find these to be mutually exclusive states of being. And the more people of color, women, gay, trans, and straight white dude "gun-toting liberals" I meet, the more I am assured I'm not the only left of center hunter or gun owner out there.

When it comes to non-gun-owners forming their opinions about guns, most of the conversation surrounds handguns, assault weapons, and any atrocity that makes them pop up in the news (over and over again). Unfortunately, stories like "woman goes hunting, brings home pheasant for dinner, has enjoyable day outside" rarely make the twenty-four-hour news cycle, so I don't blame people for only thinking of the (very, very, very) bad guys when they think about guns. My relationship to guns is different because I grew up shooting. When I hear *gun* I think of the small, handsome, wood-stock shotgun my dad got me when I was a kid. For me, that image is as innocuous as if he had given me a new baseball mitt. So, when I started getting into hunting, I was surprised at how the psychological barrier of *GUN* proved to be the source of more controversy with some conversation partners than the actual hunting part was. I'd occasionally find myself in a strange world of backward bigotry and stereotyping issuing from the left–"you have touched a gun; therefore, you probably represent these values that I will now assign to you."

Emma Marris told me this type of reaction is more common in the eastern United States than the western, where hunting is more commonplace among a broader political spectrum, but the attitude is pervasive. It even came up when I was talking to Sophie Egan about sustainable food systems, with guns being one of the biggest factors that makes people assume the political leanings of hunters. I'd asked her why hunting doesn't come up more in food writing–I've read articles about hunting in the *New York Times* and *Wall Street Journal* that mention food, but I cannot recall ever seeing an article about sustainable eating that mentioned hunting—but I didn't really need to ask; I knew the answer. And I was unpleasantly relieved to hear Egan agree with me. Think of the organizations and people that write about sustainability, food ethics, and the climate emergency. Putting aside the fantasy that politics never interfere with reporting, it's not hard to see how maybe (and I'm gonna generalize here) the left-leaning organizations that report on those topics don't want to include something they consider a right-leaning activity.

If you have a picture in your mind of a particularly repugnant character sleazing around the woods or back at camp, littering, drinking, shrugging off fair-chase ethics, or willfully ignoring game laws, you're not wrong. Those guys are out there. You can call them "slob hunters" or any derogatory term of your choosing. I'm sorry that there were or are enough of them to cement that image in your mind when you hear the word *hunter*. We hunters and target shooters do what we can to battle these poor representatives. I hope you'll remember the whole rotten apple dynamic, and that you are far more likely to hear about the actions of these particular garbage monsters than those of my friends, because when do you ever see news about people doing stuff by the book? It's outliers that you hear about the most. They're the ones who make headlines and are the best at letting their presence be known.

I'll also acknowledge that people who hunt respectfully and legally but who represent everything you despise politically also exist. People who you could not disagree with more on any political topic. It's important not to toss the baby out with the bathwater here. The hunting and conservation systems in America (and a lot of other places) work, regardless of who is paying into them. I'm sure an ultra-conservative hunter could come up with a laundry list of reasons to dislike me and my political values, but I don't care. As long as they keep pumping money into habitat systems and wildlife conservation, I'll consider us on the same team for this one thing at least. Conservation groups work tirelessly to protect habitat, water, species, and access to land regardless of the political affiliation of their membership base. Politics may occasionally come into play when some of these groups make action plans or hold fundraising events, but that doesn't invalidate the work they do for the environment.

The politization of guns isn't hard to trace, but I think there's also an assumption among people who don't know much about guns that military style assault weapons and hunting rifles are basically the same thing. They are not. I understand how it can be hard to see the difference between one sort of firearm versus another, but they aren't all the same. You can be pro-hunting and pro-gun control at the same time. Lumping hunting rifles and shotguns together with assault weapons is like comparing a Ford Taurus to an M4 Sherman tank (also a Ford product). Both will get you to Trader Joe's, but one is a tool and the other is a weapon. A gun is a dangerous tool for sure, like an axe or a car, and accidents can happen, but are very rare. And just like an axe or a car, they're also perfectly safe if you have been educated about their proper handling, have respect for what you're doing with them, and aren't an idiot or one of the aforementioned garbage monsters. I've been shooting since the fifth grade, and New York City drivers scare me way more than my shotgun ever has. Lilly Raff McCaulou, author and noted gunphobe, had to remind herself in *Call of the Mild* that "your chances of dying by accidental gunshot are 1 in 6,309. Your odds of death by cancer, on the other hand, are 1 in 7. Heart disease, 1 in 6."

If you're worried the use of guns would somehow compromise all your progressive, liberal, democratic, or whatever principles, allow me to assure you, it won't. You are responsible for being the type of hunter and gun owner you want to be. Shooting sports are for everyone. And assuming someone's political or cultural leanings because they like to hunt or shoot makes you the one creating the social and cultural division. The fabric of society is very complex, George.

I hadn't been in the state of Texas more than three hours before I set foot in a gun range. I distinctly remember feeling that was appropriate. In preparation for our pig hunt, Chris had brought several rifles to the range for me to try out. Any new avocation comes with its own furnishings, and I've always been a bit of a froth-mouthed gear junkie. Quality over quantity, of course, but searching for that ideal piece of equipment that will perfectly suit your needs and learning the tricks of a new trade is part of the fun for me. (I recently got into ham radio and boy howdy was that a rabbit hole. Related: my soul may be a good forty years older than the rest of me).

I used to shoot quite a bit when I was younger–shotguns not rifles–and I still love to shoot sporting clays. If you're unfamiliar with this sport, it's often described as "golf, in the sky, with guns." A small orange disk about the size of a bread plate, called a clay pigeon (neither made of clay nor shaped like a pigeon; discuss), is launched into the air by a machine as you stand at a station–stations are like the tees in golf. Some are thrown from towers to the right or left of you, some pop out from knee height and soar away from you in any surprise 45-degree direction, and some are launched down on their sides so they can bounce across your path on the ground. You move from station to station trying to hit all the combinations with a shotgun. Each station is meant to mimic the movements of flushing birds and jumping rabbits, but even if you never intend to go hunting (I never did), it's still a lot of fun. If you've ever been to a carnival shooting gallery, shot lasers in a theme park ride, or played *The Oregon Trail* or Nintendo's *Duck Hunt*, you're a prime candidate for having fun at a clay range.

The reason you use a shotgun for small, fast-moving targets like birds is that the spray of pellets covers a wider area. You'd never use a rifle–a bullet can travel too far, which is dangerous, and your chances of hitting a small moving target with a single projectile are too slim. My dad had gotten me a 20-gauge shotgun when I was in the fifth grade, but the recoil and loud bang scared the shit out of me, so he switched the 20 out for the even

smaller 28-gauge. In shotguns, when the number of the gauge gets bigger, the shell and subsequent firepower gets smaller. That's because a shotgun's gauge (28, 20, 12, 10) is measured by a correlation of the diameter of the hole at the end of the muzzle (its bore) and how many lead balls of that size it would take to equal one pound. So, a 12-gauge, which is probably the most common and has a bore diameter of 18.5 mm, would need twelve lead balls of that size to equal one pound. If you're going to be shooting, it's important to understand gauges; unfortunately, the systems we ended up with to categorize them are about as useful as horsepower is to understanding your car's performance, but there you have it.

Just as shotguns are classified by their gauge, rifles are classified by their caliber. First, a "bullet" is just the projectile that comes out of a round of ammunition, the whole thing together is called a cartridge or a round. The name or type of cartridge (its caliber) usually references the diameter of the rifle's bore. A .22 LR cartridge is small, tip to tip about the length of a quarter. It's used for target shooting and smaller game like squirrels and rabbits. A 243 cartridge is significantly bigger than a .22 but still on the small side of hunting rifle cartridges. It's used mostly for deer, especially the smaller deer that live in warm climates like Texas. A 308 is bigger still and can be used for deer, elk, moose, boar, or black bear.*

When I say 243 or 308, I'm referring to the cartridge or a rifle that shoots that size cartridge. People refer to their rifles using the cartridge number because that's what tells you the sort of firepower you're dealing with. A 308 round is shot out of a rifle that only shoots 308 rounds. These numbers are not interchangeable. There's a lot of outdated technicalities and nomenclature when it comes to rifle calibers, so much so that I can't even

---

*You know those old-timey, black-powder, shove-the-bullet-down-the-barrel-with-a-ramrod, Revolutionary-and-Civil-War-type rifles? They're called muzzleloaders, and people still hunt with them. You have to load them through the barrel (the muzzle), and it takes a bit of time and skill to do quickly. Most have become more streamlined and accurate since the days they were regularly in use (to the point that they often look like a modern rifle), but they don't shoot as fast or as far as centerfire rifles, so, like archery, you have to get closer to your target to hunt with one. Because of this, muzzleloaders also have their own season dates and sometimes crossbows are included in the muzzleloader season. People enjoy hunting with them because of the challenge, the gun's heritage, and the extra season dates they afford.

give you a general rule of thumb to follow, because a lot of the names have differing origins. Chalk it up to one more thing that would be a hell of a lot easier if it was based on a uniform metric system. The letters after a caliber are just as bad, most often referring to the manufacturer that first created them–for example, a .243 WIN was created by Winchester–but there are exceptions, among them LR, which stands for Long Rifle, and Creedmoor, which was actually introduced by Hornady but developed in partnership with Creedmoor Sports. Cartridge names are a chaotic wreck, annoying to those who shoot and a foreign language to others. If you'd like to know more, there are plenty of articles and videos out there on rifle calibers. Don't feel bad about needing an explainer–a lot of shooters find the minutia of rifle calibers confusing too.

When I took Hunter's Ed, I was well acquainted with the ins and outs of my 28-gauge over and under ("over and under" or "side by side" refer to the placement of the two barrels found on a break action shotgun). I was less familiar with the handling of rifles, having only ever shot a handful. I had always wanted a .22 LR so I could plink in the woods. It's a small rim-fire rifle, which means the firing pin hits the side of the round, not the back of it (that would be center-fire). It's often given to kids who have graduated from the air rifle (think Daisy's Red Ryder). Plinking is outdoor target shooting with nonstandard targets, like the tin cans and glass bottles you see in the movies. Most .22 target shooters today aim at swinging metal targets like at an old-timey shooting gallery. The name plinking is an onomatopoeia for the tinny sound that's made when you hit your target. I've never lived in a place with enough yard for me to plink safely, so when Chris told me he had a .22, I was eager for him to bring it to the range. Both so I could finally get my hands on one, and so I could feel like I was getting a quick run down the bunny slope before picking up something bigger. The other rifles he brought were a 243 and a 308. Chris was pretty sure I'd use the 308 for hog hunting but was working me up to it.

I'm svelte but gangly–my arms are skinny, and my legs basically go up to my armpits (my friend thinks I look like a black-tail jackrabbit, which, despite its name, is a hare not a rabbit). Because of my lanky build, I thought the 243 would be more comfortable for me. It's often the choice for younger or smaller hunters. To my surprise, I much preferred the heftier

308. It had a deep, thuddy recoil as opposed to the 243, which I found to be sharper and more stingy, both in report and recoil (*report* comes from the Latin word meaning "to carry" because the sound is carried a long distance). I suddenly felt like a rifle sommelier.

One question that non-gun-owners often have goes something like, "okay, I accept that you need a gun to hunt, but why do you need *so many* guns?" To which I reply, why do you need so many shoes? They do different things. Let's say you want the ability to hunt a few different types of game, but you also want to own as few guns as possible. That's three right there. A shotgun, for birds. A centerfire rifle for deer. And a rimfire rifle for small game. But that's putting a lot of pressure on those guns and the versatility they're capable of. In my fantasy gun cabinet, I'd have a handful of different shotguns, for sporting clays, upland birds, waterfowl, and turkey; a couple different centerfire rifles, one for smaller big game like deer, and one for bigger game like elk; some small game guns like a rimfire .22 long rifle and a .410 shotgun; and a black powder gun so I could hunt the muzzleloader season. So that's nine. I bet it sounds like a lot, but do you own boots, casual shoes, flip flops, sneakers, dress shoes? And do you only own *one* pair of each of those styles, or do you have two pairs of sneakers? Is one pair of boots better for hiking and the other better for snow? You're not a "gun-nut" if you own different types of hunting guns, any more than you're Imelda Marcos if you own different types of shoes. If you like hunting, and if you want to afford yourself the most opportunities to do it, then it helps to have more options.

My friend Mike still occasionally hunts with his grandfather's Winchester 30-30. It's old and heavy and packs a punch to your shoulder with its substantial recoil. He has other rifles, new ones that are lighter and sleeker. But in the same way you might have affection for your grandmother's rolling pin, hunters have the same sentiment for the rifles and shotguns passed down to them from previous generations. Imagine using grandma's rolling pin to make a pie crust from scratch then getting to share that pie with your friends and family. That pride and joy in making food for your loved ones, using a tool that your mother's mother used in the very same way, is the same pride Mike has when he shoots a deer with the family rifle and joyfully cooks up the venison for his wife, toddler, and friends.

During my visit, Mike and I were both coming up short in archery season, so a week before it was over, he wanted to start strategizing for the first day of the rifle season. We looked at maps and sketched out the area he wanted to hunt, planning to start from different sides of the mountain in hopes that any deer we didn't see or accidentally scared away would be bumped in the direction of the other. I felt like I was gearing up for D-Day.

The first morning of rifle season, I came upon a group of does in the total darkness. I could only see flashes of their eyeshine in the red light of my headlamp. I was spitting distance from them but couldn't tell if there were three or nine. They took a moment to consider what they thought they were seeing and smelling, then took off. I admonished myself for not having fortune-telling powers and spent a few moments contemplating what, had I known they were there, I would have done instead. This was a useless exercise, so, though I enjoy beating myself up about things, I snapped out of it and kept walking to what would be my designated first-light waiting point.

Mike and I both struck out on opening day, and my daily routine of walking the hills began. One day from the east, the next from the west. Sometimes I'd come back for lunch, sometimes I'd stay out all day. Sometimes I'd just go out in the morning or just in the afternoon. It wasn't always exciting, but it was always engaging enough to want to stay out longer or want to get up again at 3:30 in the morning and do it all over again.

Meditation doesn't interest me, but it's hard to argue with the meditative quality of hunting or, perhaps, its kinship with the Japanese art of forest bathing. I found it both calming and rejuvenating to walk through the woods so slowly and quietly. I couldn't daydream the way I would on a hike, because I needed to be acutely aware of everything I was seeing and hearing in the moment, and I couldn't just stomp through the woods—every step had to be slow and thoughtful. Where was I going to put my foot down? How carefully could I do it? Step, look, listen, wait, step, look, listen, wait.

One morning, I had walked all the way to the top of the mountain, across the crest, and all the way down the other side and was turning to walk back toward the house to call it a day when there he was. A young four-point buck standing barely thirty yards from me. I couldn't believe it. How was it he didn't hear me or smell me or see me? How was it I was just seeing him now? I sat down on a log and rested my arms on my knee so my rifle would be better supported. I kept expecting him to realize I was there. I waited until I had a good broadside shot. Then I took it. He stumbled a second, but, before he even had the chance to fall over, I chambered another round and shot again (after having watched my first button buck, the one I'd shot in the heart, run off, I was a little spooked and wanted to make sure this deer died just as quickly). He fell right where he was.

I sat on the log shaking for a few minutes. I was certain he was dead, but I didn't want to rush him in case he wasn't. I waited about ten minutes before slowly walking over. Even as I walked up to him, knelt down, and put my hand on his shoulder, I still couldn't believe it. He was a handsome, robust mountain deer with a wonderful streak of black hair that ran down his chest. He was young, with just three points on one antler and a single spike on the other side. His coat was getting thick for winter. I inspected him all over and was sad I could only get this close to a wild animal after it was dead. Dead or alive, connecting with a wild animal in nature will change your perspective on your small place in the world. Seeing that animal alive then knowing I was the one who made it dead grounded me and connected me to the earth and my food in a way nothing else ever had. There was a small element of pride, the same pride I imagine a gardener feels after uprooting a carrot they've grown. I didn't feel superior to the deer. I didn't feel I had conquered nature. I finally felt equal. We show dominance over nature when we remove ourselves from it, and when we think we can remove ourselves from it. Buying a cheap pork loin or dino-shaped soy nuggets at the market, that was the grotesque show of dominance over the natural world, I realized, not this. I had never set out to have some romantic experience. I wasn't trying to emulate the stories I had read from hunters past. It just happened to me. Most "ah-ha" moments make you feel smart; mine, this experience, made me feel dumb. How had I missed this before, this thing that has existed forever? I felt like I had unlocked something. I spent most of my life not understanding hunting. I felt now I'd spend the rest of my life not being able to explain it.

I gutted him, which took me longer to do by myself, then made a drag using some cord and a stick and started to pull him through the woods toward the old logging road and car path, which I knew was closer than the house. He was heavier than the deer I had shot with Fisher, but I was happy to be by myself despite the extra effort. I don't think I've ever been so contemplative in all my life. The isolation let me feel things I had missed when I had the companionship and guidance of someone more experienced. Hunters always talk about their first. The first of any animal or the first time hunting new terrain. I think maybe the two most impactful firsts are the first time you have a successful hunt, and the first time you have a successful hunt alone.

The emotions I felt were wildly divergent–it barely seemed possible to experience them at the same time. I was proud and excited but also guilty and heartsick. I had shot *that* deer, that individual who found himself in my sights. But he was also *a* deer, one of many. Carefully accounted for in the New York state wildlife management plan. And by putting the meat I took from him in my freezer, I wouldn't have to buy any from the store, farmers market, or butcher. And despite the fact that at this point in my life I was predominantly buying the "best" meat, 100 percent grass-fed or regeneratively raised, I was still happier my money was going toward preserving the deer's wild habitat, making sure the waters he drank from were clean, and paying for biologists to study wildlife diseases and changing habitats and make population surveys. I didn't think the deer would care about any of that, but in a weird way I was doing it for him, or the rest of the deer at least.

7

The Sheep Show

Tags, Points, and Lotteries

I had the privilege of being a speaker at the Wild Sheep Foundation's annual Sheep Show convention in Reno Nevada one year. Operating on a shoestring budget, I opted for an Airbnb as close to the convention center as I could get and walked to the hall each day. My schedule allowed me to spend my mornings wandering the halls, where I could peruse brochures of hunting guides, fiddle around with the latest rangefinders and gadgetry, try on top-of-the-line expedition backpacks, and discover iconic fashion like the "furkini" (it's precisely what you think it is–I was rather offended there was no option for men).

It wasn't all mountain gear and fuzzy unmentionables, though. There were as many conservation groups and state Fish and Wildlife departments with booths as there were hunting guides and outfitters. Various Fish and Game representatives were there to illustrate the hunting opportunities their state afforded and answer any questions in hopes of reeling in some new out of town hunters and their license fee dollars. The host state of Nevada was given priority placement in the lobby of the convention center where they handed out little pouches with lens cleaning cloths inside. Printed on the cloth was the 800 number for reporting poachers.

One booth I initially thought was selling some kind of ATV was actually a 501(c)(3) nonprofit that specializes in guiding hunting, camping, and fishing trips for people with disabilities in their own tricked out, off-road wheelchairs. The one on display had continuous track treads like a mini tank. I told the guy I'd like my friend to sign up, but that she didn't have great use of her hands. He didn't miss a beat and said "We can make anything work. If someone wants to be outside, we want to make it happen for them." The Rocky Mountain Goat Alliance was there too, making sure sheep weren't the only high mountain monarchs to get the spotlight.*

There was an entire second hall devoted to kids' activities. They had built an indoor pool to teach kids kayaking skills and an archery range where they could shoot at large foam disks flung into the air, like sporting clays but for bow and arrow. The pressure of social acceptability was the only thing keeping me from jumping around inside the giant fish-shaped bouncy bass castle. My favorite game in the kids' hall was an obstacle course put together by the National Bighorn Sheep Center, who had sent reps in from Wyoming. Kids had to navigate their way from start to finish by learning about and overcoming the impediments mountain sheep face throughout the year. I don't think sheep pneumonia made it into the course, but I do remember a dad shouting "Look both ways!" as his five-ish-year-old (sporting a paper sheep-horn crown) ran across the makeshift road.

There was taxidermy everywhere. Mountains were constructed along the walls to display some of the full body mounts of different sheep species. Taxidermists had brought their best pieces to show off their skill, and some guide outfits had borrowed pieces from previous clients as proof they knew how to find elusive animals. I held my own private taxidermy competition in which I secretly judged all the mounts in the convention hall. Predators like mountain lions and bobcats are incredibly difficult to mount. One teeny tiny irregularity in those forward-facing eyes–just half a centimeter off–and it's all over; you're suddenly looking at a drunk cartoon character instead of an elegant prince of the forest. This crowd didn't need to hear about conservation funding in America, so I gave my seminar

---

*Upon my return home, the stories I told about the Sheep Show led me to realize a shocking number of my friends couldn't tell the difference between wild sheep and goats. I found this to be a worrying trend that, once I started investigating the phenomenon further (I had time on my hands), extended to scores of people online using incorrect social media tags.

on the history of taxidermy. It's by far my pluckiest, most crowd-pleasing lecture, and I was honored to have some esteemed taxidermists come listen. They weren't the only wildlife artists at the show though. The hall was filled with painters and sculptors.

After one of my sessions, a tall, fit guy in his mid-fifties, with an enviable mustache, stopped to chat and tell me he was a wildlife sculptor and also a fan of the painter Carl Rungius (who I dote on in both my museum tour and taxidermy lecture). No surprise there–most wildlife artists are. He told me his name was Tim Shinabarger, and we gabbed like excited kids about our favorite animals, sculptors, and taxidermists. I knew I had met a kindred spirit. He'd been living a life I had only fantasized about; packing horses for extended backcountry hunting trips, guiding elk hunts, working as a wilderness fire fighter, and studying the great taxidermists to learn anatomy and sculpture. He had been a young sheep hunter and recounted for me how his "feet turned to hamburger" and he "bled right through [his] boots" on his first high mountain sheep hunt, all the while weaving his love of wilderness, history, and art into one connected experience. I asked to stay in touch, so he gave me his card, and when I got back home to Brooklyn, I looked up his work. I was struck by how good it was, the platonic ideal of classical wildlife bronze sculpture. I hadn't been particularly enchanted by any of the works I had seen at the show, and I suppose I was expecting something similar, but his style, compositions, and execution of anatomy and animal behavior made me gasp a small *woah* when I opened his website. His animals reminded me of 3D versions of the ones Rungius would paint, the most accurate anatomy seen through the hands of a "light" impressionist. We sent some emails back and forth and I gushed over his work. Not that he needed any affirmations from me, his sculptures can be found in several prestigious museums and galleries, including the National Museum of Wildlife Art, right alongside the Rungius paintings we both love. At shows like this, hunting and the desire to protect and admire never feel dichotomous.

Kevin Hurley, vice president of conservation for the Wild Sheep Foundation, spent forty years as a wildlife biologist before joining the Wild Sheep Foundation's team. I'd met with Kevin the year before at the Foundation's headquarters in Bozeman, Montana, to ask some basic questions about wild sheep populations, relocation programs, and disease management, but we started diving into all the foundation's current programming and before I knew it, I was eyeball deep in charts, maps, and reports. His enthusiasm for the animals was infectious. It was like listening to an eight-year-old talk about their favorite dinosaur. I couldn't write my notes fast enough. He sent me on my way with such a tall stack of books and research reports that I had to buy an extra bag at the local Goodwill to fly home with them all.

The Sheep Show is the major fundraiser for the foundation. And while it's an opportunity for all kinds of wild sheep fanatics to come together and share, the tag auction is an event highlight. When Kevin was a graduate student studying a herd of bighorn at the University of Wyoming, it was the money from these auctions that funded his research. Over the course of the weekend, thirty-four sheep hunting tags would be auctioned off, raising close to four million dollars for wild sheep conservation.

If you think hunting all North American big game is as easy as buying a license and heading into the woods to kill something, then hold onto your butts. But before we get into how permits and points work, let's take a step back to why we need them. "Carrying capacity" is how many animals a given area of land can sustain. It's an ideal, Goldilocks-style number that takes into account the area of land and the resources available to the animals that live there. Too many animals result in increased rates of disease, human-animal conflict (like car accidents and pets getting eaten), and boom-and-bust population spikes and troughs caused by starvation. Nature usually keeps overpopulation in check with predators and weather events. But if reproduction rates are steady, and barring any extreme natural (or unnatural) development, then there will be enough of a surplus in populations every year to sustain regulated hunting. In places with no other predators besides humans (and their cars), the surplus can be huge.

In the eastern United States, our big game species are primarily whitetail deer and black bear, though we do have some moose and a few small pockets of elk in the southeast (thanks to the work of the Rocky Mountain Elk Foundation). Out West and in Alaska is where you'll find America's mega megafauna, massive animals and large herds we're *very* lucky to still share the land with. Dan Flores called it the "American Serengeti," an homage to the idyllic, wildlife-filled, open spaces of Africa we all still idolize. Species like bison, elk, moose, caribou, mountain goat, mule deer, and bighorn and thin-horned sheep are all beloved and prized game species, but these animals don't have the Whac-a-Mole populations or rebound numbers that whitetail do. Deer, as you may have witnessed, are able to hack it pretty well in your backyard. Mountain goats and wild sheep are less interested in a life so close to sea-level. There are fewer of them partially because of the unique habitats they require. Therefore, the regulations that govern mountain-goat hunting must be stricter.*

The hunting rules and regulations for harvest allotment are set by each state's Fish and Game Department annually. This allows game agencies to operate on the most up-to-date data gathered by their biologists and the previous year's harvest statistics. Let's say you want to hunt mountain sheep. They are an extreme example (which is why they're fun to talk about, but keep in mind that they don't represent your average hunting or fund-generating experience), notorious for being both the most difficult tags to draw and species to hunt. Which is how they've cemented themselves as an aspirational adventure in the mythos of hunters the world over. When Steven Rinella reaches the pages dedicated to bighorn sheep hunting in his book *The Complete Guide to Hunting, Butchering, and Cooking Wild Game: Volume 1: Big Game*, he doesn't even mention the strategies or methods necessary to hunt bighorn sheep. Instead, he states that "it's a far better use of time to discuss the ways in which a hunter might get a bighorn tag." So, let's start from square one, but keep in mind each state has their own unique (and evolving) way of doing things, and I've just cherry picked a few states as examples.

Licenses Your Hunter's Ed certificate will allow you to buy a license in any state. But you must have a separate license for each state in which you wish to hunt. If you are a resident, your license will cost you much less than if you're a nonresident. For example, a hunting license costs $45 for an Alaska resident and $160 for a nonresident. In New York, it's $22 for a resident and $100 for a nonresident. Prices for everything will vary state to state.

Tags A tag is your permit to hunt one particular species in one particular state. Rinella compares licenses and tags to "an amusement park that

*Mountain-loving critters like sheep and goats are in a tough spot right now. As the climate warms, they must keep moving higher in elevation to find the food and temperatures they prefer, but eventually the mountain ends, and that's seriously bad news for those guys.

charges you an entrance fee to get through the gate [your license] and then also a fee for each individual ride you choose to take [your tags]." If you want to hunt pronghorn and elk on your trip to Montana (and man, do I hope you have a big freezer), you'll have to buy one license and two different tags. If you want to hunt pronghorn in Wyoming and elk in Colorado, that's two licenses and two tags. Tags for abundant species like whitetail deer are easy to come by. Tags for other big game animals have a shorter supply and higher demand, necessitating some clever (if a little confusing) state allocation systems.

Over-the-Counter Tags Sometimes called "non-permit tags" these are the easiest tags to come by. They're widely available, and the purchase is all you need to get going after your license. You could pick up an OTC tag at your local outdoor shop in the morning then go hunting that afternoon.

Limited-Availability Tag These are the tags you'd have to purchase when the demand exceeds the availability of the animals. A state may also choose to limit the availability of tags even if the populations of certain game species are booming. The benefit in that scenario is a higher percentage of animals getting old enough and big enough to reach a "trophy class" size. So even though the state has cut short the number of hunters who can actually hunt their land, perhaps they've increased their popularity. Why hunt elk in state A or B when you know for sure the elk in state C are under less pressure from competition and are brawny giants to boot? This might encourage people to spend more money vying for one of state C's tags. Money that goes right back to the state Fish and Game department selling those tags. It's a controversial tactic, with its own merits and drawbacks, purposefully limiting access but drawing money into that state's conservation funding pool.

To divvy up available tags fairly, states implement a lottery system or draw. All the hunters who want a tag enter the lottery, but only a few come out winners. Arizona, for example, only releases 80–100 desert bighorn tags per year, but there can be tens of thousands of applications for them. John Branch reported for the *New York Times* that "in Montana, 19,439 applications were submitted by state residents in 2015 for bighorn sheep; licenses went to 111, a success rate of about 1 in 200."

Also, just because 111 tags were doled out, that doesn't mean 111 sheep were shot. The number of successful hunters that year was 107. Success rates vary based on species and location. Some states cap the number of tags allotted to out-of-state folks so their residents can have a home field advantage. Resident or not, you can apply for these tags for years on end before you get one, or you might never get one at all. Because of that, bonus and preference points are awarded to people who apply for the same tag more than once, in order to encourage them to continue applying each year.

Preference Points These points accrue each time you apply for a tag and get skunked. If you apply for a tag one year and don't get it, you get a point instead. The more points you have, the more likely you are to draw a tag the next season. A maximum point holder is someone who has accrued (you guessed it) the most points. States will gather up all their max point holders and set aside tags to be drawn just for them–their loyalty as repeat applicants has earned them the privilege. The following tier of tags go to the next highest point holders and so on. Points are not transferable. If you have six points toward a Montana bighorn sheep tag, that's the only thing those points apply to. You can't use them in another state or for another species.

Bonus Points These also encourage you to apply for the same tag annually. If you put in for a tag for the first time, your chances of drawing it are slim, like under-one-percent slim. But after a few unsuccessful draws (which means a few unsuccessful years) you might be eligible to start accruing bonus points, which will allow you one additional entry for every year you apply, effectively doubling the number of "tickets" you're buying for the lotto. Some states will even square your number of bonus points. In *Wild Sheep* magazine (shout out to all the wild sheep-specific magazines out there . . . oh that's the only one?), Craig Boddington wrote that he had applied for a Rocky Mountain goat tag in Montana for twenty-nine years and an Arizona desert sheep tag for thirty years without getting one. In Branch's *New York Times* article, he quoted one man, George Dieruf, as saying, "I've been putting in for a sheep tag since I was twelve years old and I've never gotten one, and I'm sixty-seven." Some states have other things you can do besides entering the lottery to gain points. Arizona offers one bonus point to adult hunters who complete the AZ hunter safety course and participate in a field test.

In order to weed out the daydream hunters who, as my father would put it, "have eyes bigger than their stomach," some states will require you to pay for the tag you want upfront. That way they've already got the money if you win the draw. If you don't win, then they'll refund you. Rinella plays out the math for us and explains that "if a hunter applies for a bighorn sheep tag in every state that allows nonresident applicants, he'll have temporary expenditures totaling around $8,000 annually and permanent expenditures of about $600 annually even if he doesn't draw a tag."

As I learned more about these convoluted systems, I went ham on a tag-application spree. Tag applications and points weren't something I had much experience with as an East Coast deer and turkey hunter, so I wanted to play around with other state's online systems and application procedures. I spent a couple hundred dollars entering lotteries in Washington, Wyoming, Montana, Colorado, and Arizona. I won a tag for a pronghorn doe in Wyoming but knew I'd be hard pressed to make that hunt happen that year. I didn't feel bad about the money, though–I was happy to "donate" to all those Fish and Game agencies.

The thing that annoyed me most was how different each state's website was. It's reasonable for states to handle their point and draw systems differently based on their individual needs, but I estimate a lot of money is being left on the table by not having uniform visuals and organization from state to state, which would make tag applications more user friendly and therefore more enticing to apply for. The complexity of the systems and the degree to which they vary by state has been the impetus for the founding of a few data aggregate websites that will, for a fee, help you apply for tags in areas you're more likely to win, or to annually enter draws for you, so you don't have to think about it.

Auctions and Raffles Auctioning tags can be a touch controversial. Some people think they break the North American Model of Conservation rule that hunting should be democratic and open to everyone, not just the elite, but I don't have a problem with them. They produce staggering amounts of revenue for conservation organizations and programs. It would be problematic if auctions were the only way to acquire tags, but auction and raffle tags represent the minority of tags available, despite being responsible for the majority of funding raised by tags for some conservation programs. The difference between the money generated by Average J. Hunter and Dr. Big Auction Winner is well worth saving a tag for the highest bidder.

Let's say (in an unrealistic waking fantasy) that I put my name in the hat for a Montana bighorn sheep tag and won it. I'd have to pay $1,325 to the Montana Fish, Wildlife and Parks Department. That's $10 for a Conservation License, $15 for the Base Hunting License, a $50 nonrefundable application fee, and the $1,250 for the winning tag (winning tags are just $125 to MT residents). If I didn't win the tag, I'd still have to pay the $75 for the other three licenses and fees. Meanwhile the record auction price for a Montana bighorn tag is $480,000. Ten percent of that went back to the Wild Sheep Foundation, which auctioned off the tag. Ninety percent went right to Montana's wild sheep conservation programming.

If sheep tags were only saved for Average J. Hunters, there might be fewer tags ultimately available to them. The massive injection of funds produced by big auction winners fund 74 percent of wild sheep conservation projects. Montana Fish, Wildlife and Parks generates around $120,000 from the sale of licenses and tags to be used for sheep conservation programs. Gray Thornton, president and CEO of the Wild Sheep Foundation, says that amount "wouldn't pay for one biologist and a truck." The support of these auction spenders is how herd-health programs and relocation efforts get funded. It's not cheap or easy to have a vet perform a battery of tests on a bunch of wild animals and then fly them by helicopter to populate herds in new locations. You have to really, really like wild sheep to bother.

8

Talking Turkey

New Hunters and Meeting People Where They Are

"The development of the cannon net … ultimately spawned the boom in wild turkey populations nationwide."

This is perhaps my favorite line in Matt Lindler's "Orchestrating a Comeback," an essay about what it took to bring wild turkey back from the brink (I knew Lindler must be a decent guy based on the fact he didn't italicize the word *boom* in the above quote the way I would have). Just like so many other species, turkey had been nearly wiped out by the early 1900s. Some estimates put their numbers at only 100–200k. I like old Benny Frank's partial jest that the wild turkey would make a better national bird than the bald eagle. Bald eagles are sexy looking, but they're basically giant hunky thieving, scavenging, garbage seagulls. Turkey, on the other hand, are proud and wise. And they are beloved quarry to the hunters who pursue them. Every turkey hunter I've ever talked to or heard from is a total turkey-obsessed fanatic. People don't love athletic teams as much as turkey hunters love turkey. Even if you couldn't care less about these ungainly birds, they make for a great case study of why hunters fall in love with particular species and why it's good that they do.

In 1973, pissed that ducks and bison and other game animals were getting all the conservation attention, Tom Rogers founded a community of turkey addicts called the National Wild Turkey Federation. At the time, turkey populations had risen a bit—numbering just over one million, but that wasn't good enough for our man Tom. Rogers saw that state conservation programing was lacking when it came to turkey-specific initiatives and decided the federation would have to become the *wing* of America's wild turkey restoration, management, and conservation for the future. I wanted to know more, so I called up Matt Linder, author of the boom quote and director of government affairs for the NWTF, who kindly took a break from his work while I kept him on the line with probing questions about cannon nets and turkey boxes.

First of all, there's a Wild Turkey Technical Committee. Representatives from almost all fifty states get together to discuss major turkey-related issues in their state and regions, and game plans are *hatched* for the future. These meetings are not held in a Dr. No–style subterranean lair, but they're still pretty cool. Mark Hatfield, the national director of conservation services for NWTF takes an appropriate amount of pride in the program, stating that "over the course of more than four decades, the committee has been instrumental in focusing research, policy and management techniques to the benefit of wild turkey and, indirectly, many other species."

Multiple initiatives were tested out over the years to aid national wild turkey population recovery. One of the earliest methods tested was to hatch wild turkey eggs in a protected space, offer them safety for the most vulnerable first two weeks of their lives, then release them into the wild. Sort of like stocking trout. But even with minimal handling, those birds didn't have what it takes to hack it in the woods. The most effective method, Lindler explained to me, "was trapping live wild birds from areas with dense wild turkey populations and relocating them to areas of good habitat but with few or no turkeys." But live trapping a turkey is not like shooting a tranquilizer dart into the giant butt of a rhinoceros.

Eventually, Team Turkey learned to employ a slightly modified version of a waterfowl cannon net. These types of nets are used most often by biologists who have to trap a bunch of migrating birds like ducks at once so they can band them. The birds are baited into an area with corn or another tasty grain, and when there are enough concentrated together, the cannon goes off and the net expands out over them, with the bottom edges weighted down to keep clever individuals from sneaking out the sides. The cannons look like World War II–era M2 mortars and are powered by compressed air. Once the biologists get a hold of the birds, they can measure, weigh, and band them before putting them in a specialized transport box and sending them to their new home. NWTF is *the* supplier of turkey transport boxes to state game agencies.

Caption TK

There are five major subspecies of turkey:

Eastern The most abundant and widely distributed. From the middle of the country to the East Coast.

Rio Grande Found in Texas and the smack center of the country.

Merriam's These are your western state mountain birds.

Gould's Found in a few areas of Nevada and Arizona. But mostly Mexico.

Osceola Only in Florida!

Ocellated Do not live in the United States. They're primarily found in the Yucatan Peninsula of Mexico but also parts of Belize and Guatemala. (They're resplendent; Google if you haven't seen one.)

The Gould's turkey was nearly extinct in the United States when NWTF set their sights on repopulating them in the Southwest. That meant working with Mexican game agencies to capture enough birds to bring over the border. By this time, the NWTF had perfected their formula for repopulation. With one or two toms and fifteen-ish hens, they could see populations rebound in as little as five years. But while planning the international operation, team leaders had to dust off older models of the turkey cannon–to comply with Mexican regulations prohibiting the type of propellent used in the cannons at the time–and there was a bit of a re-learning curve while team members deciphered how to use the outdated drop nets. Local farmers were paid if they wanted to help catch turkey. Lindler told me some of these farmers would walk five to seven miles one way to get bags of turkey-bait grain then carry it over their shoulder for the long walk home. One farmer, upon realizing the compensation he was being offered was per-bird and not total, shed a tear of surprised joy. When the birds arrived in the United States, they spent some time in quarantine before being released.

The precolonial population high for turkey is estimated to be around ten million. By 2004, thanks to the efforts of the NWTF, we hit a turkey population estimate of seven million. And it all started with a hunter passionate about protecting his favorite quarry.

The first spring I met him, Mike invited me to his place in the Catskills to spend a week turkey hunting. Even after I had decided to take up hunting, I had missed two turkey seasons because of my schedule and the Covid-19 pandemic, including one trip to Washington State to hunt with some friends I'd met at the Sheep Show–by the time I got up to the Catskills, I could barely contain myself. I honestly didn't think I'd get one (and I was right), but I was just excited to finally be getting out there after them.

Turkeys are wary, even for a prey animal, and that can make them extra difficult to hunt. Even though they have a poorly developed sense of smell, they have pretty damn good hearing for not having any external sound funnels like the big ears of a deer or fox, but their biggest defense is incredible eyesight. If you can see a turkey, it can see you. That means moving slower than cold molasses if one is nearby, and it's one of the reasons the pressure of turkey hunting can be so exciting. Another big part of turkey mania, and what drives turkey hunters to obsession, is that you can talk to them. It brings a whole new level of difficulty and reward to the hunter who takes the time to learn their language (you'll find the same rabid obsession among elk hunters because elk are also extra vocal during their mating season). Imagine you're using two hands to work the turkey calling device of your choice, and you see a bird start walking toward you. Now you have to find the right time to put your call down and pick up your shotgun, all while moving slowly enough that you won't attract attention.

When you call turkey, the vast majority of the time, you'll be making the calls of a hen, trying to lure a horny tom toward you. Unless you're on a wide stretch of private land, making the calls of a rival tom can be dangerous, because another hunter might mistake you for the real thing. It's the same reason hunters are warned to never wear anything red, white, or blue into spring turkey woods, as those are the bright colors found on the flushed head of a strutting tom–not a great idea to either resemble or sound like something people are looking to shoot at. So, for safety's sake, it's generally best to mimic the sounds of a hen.

The tricky part though, is that that's not how turkeys do things. Normally, a big hunky tom will strut his stuff and give a supercilious gobble in order to bring the local hens to him. So, right away you're asking this wild creature to behave a little against its nature. One strategy turkey hunters can employ to make up for this slightly unnatural behavior, if their turkey vocabulary and accents are robust enough, is the use of daytime drama storytelling.

Imagine our big mature tom turkey is a guy in his apartment (Tom). Our hunter is the guy in the apartment next door (Hunter). Hunter cups his hands and starts shouting at the wall between them in a high-pitched feminine voice. "Oh, I'm sooooo lonely. I wish some big, strong man was here to keep me company."

Hunter has Tom's attention now, but Tom's a little suspicious and waiting to hear more from this mystery woman.

"Maybe I'll take a nice hot bath. Guess I had better take off these itchy clothes first. Oh, What's that? A knock at the door?"

Tom is in total suspense now. Hunter changes up his voice to that of a cracking, scrawny young sophomoric dude (a jake, in the turkey world). "Oh, hello madam, did you order a pizza?"

"Why yes! Oh, but you must think I'm so silly, answering the door with no clothes on. Would you please come in and help me find my robe?"

Tom can't take anymore and rolls up his sleeves. "Like hell he will! *I'll* help her find her robe, but first I'm gonna beat the crap outta that guy."

And there you have it, the subtle and delicate art of turkey calling.

I spent the winter before my hunt with Mike practicing my hen turkey calls. The tool that makes the sound is also generally referred to as a call, and there are several different types. Special, Inspector Gadget-style vests are made for turkey hunters, with pockets all over them just for the arsenal of calls one hunter might want to have at their disposal in the field. Slate calls are a little circle of chalkboard you scratch at with a wooden dowel (some prefer glass to slate because it's easier to use when wet). Push-button calls (my first type of turkey call, chosen for ease of use), consist of a wooden dowel you push through a small hole in a tiny box with no lid; the dowel has a little metal claw that scratches the sides of the box as you push it. There's also a box call, one of the most traditional styles. Picture a long thin box, this time with a handled lid across the long top that is nailed in one place at an end. The pivot point allows you to slide and waggle the lid from side to side, scraping the edges of the box to produce your hen calls.

Some of these wooden calls are handmade by skilled craftsmen, and while a mass-produced turkey call can go for sixty dollars, the handmade ones easily fetch prices in the hundreds. Call collecting has a similar market to the collecting of handmade duck decoys, a once common American craft now overshadowed by the cheap (and very often not so cheap) mass produced. I have some duck decoys, because I think they look beautiful, but I have yet to dip my toe into the world of master-class animal calls. The differences between the sound of the store bought and the custom made might be lost on me as long as my turkey calling skills remain at the novice level–the difference between a piano tutor's little upright and

Van Cliburn's favorite Steinway—but there's no denying their aesthetic or cultural value even if you don't know how to use one.

Lastly, there's a mouth call or diaphragm call, which is flat and about the same diameter and shape of a round cracker cut in half. To make it squeak correctly, you have to place it in precisely the right spot against the roof of your mouth, with your tongue in a particular position to force air through layers of folded latex in the center of the call. I wish I could describe it better, but I can barely use one. If you can make it work, though, a mouth call is ideal. It's small, and you get to have your hands free. Most of my mouth call practice sessions were spent retrieving said call after I had spit it across the living room. When I was practicing with the slate call, I tried to be respectful of my neighbors and only *yelp*, *purr*, *cluck*, or *cutt* when I would go outside for a walk, but I liked to practice my mouth call inside—I wanted people nearby if I was going to choke on the damn thing.

Mike and I went hunting together twice, and he did all the calling. That way I could just focus on my shot if I had an opportunity. Like hog hunting in the blind, I was happy to have one less thing to worry about on my first outing. We didn't see anything though. On our third hunt, we met up with Mike's pal Bob, who was also his Hunter's Ed instructor. Bob is a turkey caller and hunter of some renown in the Catskills and won the New York Backcountry Hunters and Anglers turkey calling championship in 2020. I could not have been in better hands.

They both brought their own shotguns but were far more excited about calling a bird in for me than themselves. It made me think of my dad giving my niece ice cream for the first time, and how tickled he was to be there for that moment and have that privilege.

We hiked into one of Bob's secret spots in total darkness and set ourselves up against some trees that he strategically picked out for us. We sat there silently for a while, letting the memory of the mild commotion we made coming in fade away before Bob started calling. He didn't want to bring any toms in at first, it was too early and still dark, but he wanted to know where they were. And all of a sudden, he was talking with three different birds. I could hear their responses coming from all over the woods. Eventually one flew down from his perch right behind me. It was either a turkey or a small B17 Flying Fortress—I could not believe the sound he made, like a jet pack sputtering out as it reduced thrust so its pilot could safely put feet on the ground. It was one of the most exciting moments I've ever had in the woods, and I never even saw him, because he stayed someplace behind me.

There was a lot of action that morning. Two of the turkeys Bob was talking to were jakes, but one was a big mature tom. I only caught one glimpse of him through the trees, all puffed up and strutting. Bob later told me he'd had a clear shot at him but didn't take it, hoping he could bring him in close enough for me. I was touched by the sacrifice but would have been just as excited to have been a part of Bob's successful hunt as my own. I didn't get a turkey that season. But I did get hooked.

For dinner that night, we ate some turkey Mike had shot and processed earlier in the season. He showed me a photo of its crop (where you'll find the most recent foods a bird's been eating), and it held at least three different types of bright berries and a variety of greens. A fresh and vibrant salad if I've ever seen one. I tried to think of what the inside of a Butterball turkey's crop would look like. Mike's bird was better tasting than any turkey I had ever bought, and I was thrilled that we were all talking about it. The turkey was special, it wasn't just a part of a meal, it was the reason for the meal. I've never heard anyone pay as much respect to, or gain so much joy from, anything store bought, even at Thanksgiving.

Unfortunately, turkey populations have back slid since their high in 2004, and we've lost around a million birds, leaving the US population hovering around six million. There's a combination of factors leading the decline, but the big three are also problematic issues for species other than turkey. First, people are afraid of or perhaps don't understand prescribed burns, so it can be hard for them to get the green light in some skeptical regions. But the type of plant life that grows after a fire, called "early successional" growth, is vital to turkeys for nesting, eating, and roosting. Other animals need it too for similar reasons. Second, there are more small predators around. These days, trappers and small game hunters are fewer, which has led to an increase in populations of species like raccoons, coyotes, and mustelids (weasels, marten, mink), and those guys all prey on young turkeys. If the birds can survive for two weeks after hatching, they have a much better shot at survival, but that's gotten harder with more small predators around. Third, urbanization and deforestation. This is pretty much every animal's biggest problem everywhere. More sprawl equals less habitat. *You can't save any animal if you can't save where it lives.*

The issue of habitat loss was so paramount to NWTF that they launched their 2012 Save the Habitat Save the Hunt campaign in direct response. The organization estimates that the acres of wildlife habitat lost every day add up to land the size of Yellowstone annually. They see their R3 efforts deeply tied to the land as well. If they can recruit more hunters, that means they've recruited more habitat advocates–people who would rather see trees and wild animals than new development. They've had success teaming up with Fish and Wildlife services and other conservation organizations to "meet people in their space" as Matt Lindler puts it. "You have to go to them." And by "go to them" he means events like a wild turkey and microbrew pairing night at a local tap house. It's funny how quickly our perception of something changes based on the context. I know a lot of people who would turn their noses up at the thought of hunting or hunters. But ask those very same people if they'd like to try some local wild game with a beer and they don't have to think twice.

This method of using game meat as a tasty gateway to talk about hunting is called venison diplomacy. Hunters are proud of and generous with the meat they've harvested from their exploits. I've never hung out at a friend's house and had them go into their freezer to gift me some ground beef they bought at Trader Joe's, but, on many an occasion, I have left the home of a hunter with a frozen, hard-earned treasure of venison, turkey, moose, and even Dall sheep. It's not much different than proudly stating to your dining companions "I grew these tomatoes." It's sweat equity you're choosing to give away to someone. How many cookbooks say that food is love?

The National Deer Association's Field to Fork program had massive success starting conversations at farmers markets while doling out samples of venison sausage. Programs like that go beyond starting conversations about hunting–they find new places to have those conversations. Farmers markets are food-focused venues where all sorts of people congregate looking for local, seasonal, environmentally friendly fare. Bringing the benefits of hunting to non-hunters in spaces like this with nonconfrontational activities is an important way to expose more people to the lifestyle.

Sharing hunt photos on social media, along with recipes or resulting meals, is also a form of venison diplomacy. But it gets a mixed response. I've talked about grip 'n grins before but the conversation doesn't ever seem to stop. I often see the backward approach go over well: a photo of someone's finished meal ready to be eaten, a photo of the ingredients and cooking process, and, last, the grip 'n grin. Because before you can cook you have to harvest. Some folks just show off their meals and talk about the hunt in the caption. That works too. But I think it's a shame not to show the whole animal, a fundamental disagreement between the proud hunter and the offended viewer of what, exactly, respect is. To me, respect is seeing the animal as a whole, as an animal, not just the ingredients it will later yield. This is why it can be so hard for hunters to communicate with non- or anti-hunters. We are speaking different languages.

In a similar visibility vein, one of the most important events in the last decade of hunting's history probably wasn't the passage of any legislation but instead the 2017 addition of the show *Meat Eater* to Netflix. Each episode follows author Steven Rinella on different hunting and cooking adventures around the United States as well as some occasional international trips (he's the guy who took Tim Ferriss on that doe hunt in 2011). The show is beautifully produced by Zero Point Zero Production, the same team behind numerous travel and cooking shows, including (saint) Anthony Bourdain's *No Reservations*, *The Layover*, and *Parts Unknown*. *Meat Eater* actually premiered in 2012 on the Sportsman Channel, a cable and satellite-only option, but broadcasting solely on Sportsman meant, 1) the show was only being seen by people who had cable (and therefore a television), and 2) that it existed in a bubble of hunting and fishing programing that you'd have to already really love hunting and fishing to watch and pay for. A debut on Netflix meant the show was instantly in a mix of diverse programming and visible to a younger generation of people (those who have renounced their $80/month television cable package in favor of eight $10 streaming services).

Despite Rinella's impressive body of work, it's the fact that his show is on Netflix (for the moment at least) that takes him out of the celebrity hunter in-crowd space and plops him in the hunting ambassador role. Rinella isn't the only hunting role model with a television show, but his is the most accidentally accessible to those not specifically searching for hunting content. I kid you not, my friend texted me while I was writing this and asked me if I had ever seen *Meat Eater*. It had just popped up on his Netflix feed. He and his wife are watching it as I write and neither of them hunt, but both love to cook. Perhaps the almighty Netflix algorithm recommended it to them based on previous viewing habits centered around nature, food, and travel. It reminds me of the furor around *The Omnivore's Dilemma* and how it inspired the glut of food-for-thought (or thoughts about food) books that followed it. Nine years prior to Pollan's book, Richard Nelson's *Heart and Blood*, an "it's complicated" love letter to deer, said just as much just as persuasively about the cycles of food in America as Pollan did—but who picks up a 360-page book about deer if they're not a total deer freak or natural history nerd, even if the lesson is broader than the topic? *Food* is an easier pill to push than *the social and ecological implications of one particular species in American history and culture, and how it pertains to you, regardless of what you eat.*

If you're not a hunter it's a big commitment to pick up and purchase a book about hunting. (Picture a sudden and horrific realization washing over my face as I type.) But a twenty-two-minute episode of television on a platform you already pay for or steal from your friend? Sure, give it a whirl. Many incidental *Meat Eater* viewers credit Rinella with changing their minds about what hunting is and, perhaps more importantly, who hunters are, as well as kindling their own newfound interest in hunting. He makes a great fit for Zero Point Zero Production's style of docudrama and is an appealing character to viewers in his similarities to Bourdain. They're both talented authors capable of translating their feelings to the small screen, and both are interested in authenticity, curious about new experiences, and eager for hard-earned satisfaction. Even though there are hunting and outdoor recreation advocacy groups that spend big money to reach out to new people, it's tough to beat entertaining advertisements for the hunting lifestyle that come right into your home when you aren't even searching for them.

Recruitment is a hot issue in the hunting community. Not all hunters are in favor of growing our numbers, and I don't hold their concerns against them. First of all, the more hunters in the woods the more "pressure" there is on the animals. That pressure makes animals extra wary and causes them to move through spaces erratically to avoid predation. Second, if you're enjoying the solitude of nature and the illusion of a quiet, human-less, Eden-like forest, you're probably not psyched to see a bunch of competition walking around. And finally, huntable lands can only sustain so many animals and so many hunters. In order to support the number of hunters we had as a nation in the 1980s (the height of hunting license sales) we'd need to expand our public lands and find better ways to connect private landowners with hunters. A hunter used to be able to knock on a door and ask permission to hunt someone's property, but with expanding private lands owned by liability-shy corporations instead of Ma 'n Pa Farmer, that becomes increasingly difficult. If private land shrinks further from being accessible to local hunters, that puts more pressure on animals that live on public land as well as the land itself. It also changes what could be a local food source into one that needs to be sought farther from home.

Ultimately though, we have to think about the longevity of hunting itself, and a big part of that is its public perception. More new hunters, and more hunters who represent our country's shifting demographics as well as our political spectrum, mean there's more people to support the lifestyle both politically and economically. These new, young, diverse hunters can also represent the lifestyle to people who may have preconceived notions about what hunting is and who hunters are. Growing the numbers of new hunting allies is as important (maybe more important) as creating new hunters. Hunters don't need to become the media darlings of the outdoor world, but how people view us matters. When hunting-related issues end up in legislation, it's mostly non-hunters who cast votes on them, because more non-hunters exist than hunters. Today, only about 5 percent of Americans hunt. What percentage of the population do you think knows the first thing about hunting regulations, the ecology economy, or wildlife management strategy and funding structure in the United States? If all you know about hunting and hunters is what you've seen in the movies and on the news, then you might just think "fuck those guys, they shot Bambi's mom" without ever understanding that hunters did no such thing (those guys were poachers, remember). It's deer hunters who spend the most time and money trying to keep deer protected from habitat loss, car strikes, and disease—but that's rarely what's reported. As John Burroughs wrote in *Camping and Tramping with Roosevelt*, "Is it anything more than ordinary newspaper enterprise to turn a mouse into a moose?"

When writing about new hunters for periodicals, it seems authors specifically choose or avoid the term *hipster*. The *New York Times* referred to millennials as representing a "new breed of hunter" in 2019. The *Wall Street Journal* called them hipsters the same year. Other articles do the dance of coded language using combinations of words like *young*, *millennial*, *urban*, *liberal*, and *locavore*. As far as stereotypical hipster connotations go, I don't really see a problem. What do we think of when we picture the archetypal urban hipster? Someone who makes their own pickles, rides a bike everywhere, buys vintage clothes, and pays extra for artisanal bread or a coffee table made by a local craftsman out of reclaimed wood. What's the problem? All I see here is an appreciation for food, skill, craft, and a more environmentally sustainable lifestyle. It's not too difficult to see why people who value those things might progress from farmers market meat to the next and final rung of the sustainable meat ladder. Early adopters of an ancient craft. (Mmmm meat ladder.)

My only problem with the term *hipster* is the allusion to the temporary. If something is hip, it's in fashion. But fashions change. Hunting is not a fashion, it's a (life)style. Style tends to be lasting. You might trim the curled ends off your handlebar moustache one day or pull your mom jeans down from under your armpits, but I doubt you'll want to go back to grocery store beef once you've tasted and shared your own wild-harvested venison.

Hipster also generally refers to a specific age range, which can limit its usefulness. The clunky and slightly medical sounding *late onset adult hunter* was the first go-to term for any new hunter who came to hunting of their own volition regardless of their age. *New hunter*, *new adult hunter*, and *locavore hunter* thankfully appear to be catching up to replace it. I do like the mention of adulthood if there's no obvious visage to accompany the term. I'm glad my dad taught me how to shoot when I was a kid, but it was important for me to come to hunting on my own, though I am sorry it took me so long. As much as I now bemoan the years of good food and outdoor adventures I've missed, I like that I can say I chose this lifestyle after putting thirty years of thought into it—it wasn't just something I accepted as a kid because that's what the grown-ups taught me. I might be deeply envious of the life-long hunter, but I do find satisfaction in having gotten to this point on my own and choosing it for myself. Hunters are proud to be hunters. It's not something you keep doing your whole life or pick up later in your life if you're not.

The hunting world is mostly male, mostly straight, mostly cisgender, and mostly white. I'm sure you're shocked. I suppose this is the part where I tell you I'm a transman. I didn't initially want to bring it up because I find it irrelevant to the points I'm making about hunting, conservation, and sustainability, but I recognize that it might make a difference to some folks. I've always felt comfortable speaking up for animals. I understood them before I understood myself. My love of animals was a perfect excuse for me to avoid people when I was younger. It was hard to be a boy around other kids and grown-ups when they didn't see me that way. Animals just saw me as a person, so I was always able to feel like myself around them. I still feel the most like myself around animals, but I think that's just because I'm happiest around them.

Because I look like just another white guy, I've had an easier ride than so many members of the LGBT+ community and find myself confident and comfortable in almost any space (exceptions include anyplace you might hear the words *crystals* and *cleanse* in the same sentence), including ones that have guns. I don't dwell on my identity as a transman when I discuss hunting, because I'm not exactly "visible." I'm not a woman, and I'm not a person of color. When I'm in a room full of mostly white, mostly male, and often conservative dudes, which happens regularly on some of my speaking tours, I look just like the rest of them (though I probably have a better head of hair—it's as thick as otter fur—for now). One time, in Texas, I did have to hand over my driver's license, which still had an F on it, to the attendant at a gun range before renting a practice shooting lane, and I got quite the dagger-eyed glare in return. I can't say that sweaty rush of defensive adrenaline was the most pleasant feeling, but it's the closest I've ever come to trouble. I'm lucky.

Fortunately for everyone, there are now hunting organizations dedicated to women, people of color, queer folks, and the disabled that all offer community and opportunity to engage with one's peer group, and online resources have made affinity groups and communities much easier to find. Since 2011, women have represented the fastest growing group of new hunters in America, now followed closely by Hispanic people (to be clear there have always been women and people of color who hunt—they just weren't as visible in mainstream media). An influx of new hunters brings with them a demographic change (teeny tiny though it may be), often representing an injection of youth and diversity. If hunting is going to have a broad base of support and adoption into the future, then, as the demographics of the country change, the demographics of hunting should change with them.

The journey of a new hunter who is a woman or a person of color needn't be different from that of any new hunter. Today's dedicated organizations offer environments geared toward curious rookies who might not feel comfortable in hunting spaces filled with a bunch of white dudes. The most widely known groups devoted to supporting women in hunting are BOW and Artemis. BOW stands for Becoming an Outdoors Woman, and their programs are hosted by state environmental conservation and natural resource departments all over the country. As the name suggests, they aren't just about hunting but an array of outdoor recreation and safety. Artemis is the National Wildlife Federation's sportswoman's network. They have a podcast and newsletter and provide support to women in outdoor leadership roles.

For people of color concerned about gun ownership, I'd recommend the book *The Second* by Carol Anderson, a professor and historian. It isn't meant to have a pro- or anti-gun agenda but to instead focus on the historical context of gun ownership by African Americans past and present. I'd also recommend reaching out to organizations like the Minority Outdoor

Alliance, Hunters of Color, and Outdoor Afro, which might have events in your area or be able to put you in touch with someone who has personal experience in the hunting space. HECHO, Hispanics Enjoying Camping, Hunting, and the Outdoors, is another good resource. Like Artemis, it's a partner of the National Wildlife Federation.

I would recommend any new hunter join Backcountry Hunters and Anglers–I've found it to be the most social community of outdoor enthusiasts and therefore a great way to meet people. A friend of mine in Colorado just joined and she doesn't even hunt or fish! . . . yet. You can (and should) collect memberships from other groups as you find your community and the specific lands and animals you feel most connected to. The programing you can find through the National Wildlife Federation and the National Deer Association's Field to Fork classes are also wonderfully welcoming to all newcomers, and more programs geared toward new hunters are popping up all over the country. If gun ownership (and this goes for everybody) feels like too much of a deep end to dive into right away, then consider starting with archery. The hunting community is expanding, and it's becoming fantastically accessible. There's a lot to be excited about. As The Minority Outdoor Alliance and Hunters of Color say, "the outdoors are for everyone."

9

Loved to Death

How Loving Animals Can Get in the Way of Helping Them

Charismatic megafauna (perhaps the best term in the world) broadly refers to animals that have mass appeal. And this appeal, like the preference for the bald eagle over the turkey, pretty much just comes down to looks. The term is a bit redundant, I think, as most megafauna (big animals) are pretty charismatic, but the charisma part of the term refers specifically to animals who are deemed cute or sexy enough to care about. (I'm sure my fellow naturalists would prefer I start using *majestic* or something, instead of *sexy*, but you get it.) Conservation organizations use this bias to help muster up donations, because while everyone has seen pandas and thinks they're cute, it's harder to generate the same level of financial enthusiasm for the Titicaca scrotum frog. Freshwater muscles are in dire straits in the United States but rarely receive the teary-eyed, plea-for-help treatment that wolves benefit from, which is too bad, considering muscles clean the water so many species depend on, and wolves are actually doing really well these days. There's a lovely little UK-based organization called The Ugly Animal Preservation Society that plays with this convention, using a lot of humor (or humour, in deference to our English friends) to educate the public about such oft-sidelined, less-than-conventionally-gorgeous creatures.

It's not all bad to pin your entire conservation platform on the shoulders of one species, though. Let's say you want to protect caribou; well, we know them to cover the most ground of any terrestrial migrating species, so if you want to protect the caribou, then you have to protect all the land they cross—and that'll be a boon to all the species that share their habitat. Maybe you find "the rainforest" a generic and uninspiring target for your twenty-dollar donation, but you sure do like those cute, silly, orange orangutans. Saving one means you save the other. (Also, stop buying stuff with palm oil in it.) Hunters participate in this conservation cycle anytime they geek out on one species. If you're crazy for ducks, then you have to ensure those ducks have places to breed, nest, feed, and rest all along their massive migration routes. Turkey fanatics must maintain forest health and repair flood zones if they want the population to grow. Elk obsessives most efficiently protect the species by buying up more land for them to live on and safeguarding their migration routes and winter range habitats. These efforts don't just help the ducks, turkey, and elk—they help every animal who shares the landscape with them.

"Animal lover" was an easy title to give myself when I was younger, but as I got older, I realized I had to make choices for that name to mean anything—I had to try to live an as environmentally and animal friendly life as I could. Which doesn't look like what I thought it would when I started and can also be exhausting, confusing, and depressing. I spent years slowly letting light into the aperture of my judgement, and on the back of each new understanding came some related topic I had to examine next. It wasn't until I stopped personalizing sources and started to look at what was proven to be best for the environment—not what best supported the answers I wanted most—that I was able to start making meaningful choices. In my younger days, I was all for plant-based meat alternatives and banning trophy imports and hunting, but when I stopped to inspect the consequences of those actions, it was plain these measures weren't the great salves to our environmental woes that I'd been promised. Wanting to look at everything more closely made me realize I had to step back to see more instead.

It was easy for kid-Brant to look at the grip 'n grin photo of a hunter with a dead animal and say "Ah ha! You are the bad guy. I can see you've killed that deer." But what I didn't see in the picture was how the industry and funding systems built around that dead deer was ultimately beneficial to the environment and the species overall. Now, show me a picture of a family vacationing in Venice on a couple gondolas, and I can say the same thing. "Oh, *you* must be the bad guy. First you had four kids (What is this 1863? Do you need them to till your fields? Maybe you're worried a few are gonna die from consumption and you need to have spares?) Then you traveled overseas in a fossil-fuel-guzzling aircraft, and out of all the money you're spending, none of it supports the environment. In fact, your transatlantic flight is contributing to the accelerated sea level rise responsible for sinking the very city you're visiting."

I'm not going to stop traveling, I love to travel. But now I find greater cause to wag my finger at my wanderlust than my hunting habits. Zooming out was key. What's more important: the visibility of people's actions or the sometimes-invisible consequences of them? I thought the easy answers were more likely to be the right ones, but that went right out the window when I realized there are no easy answers.

There's a difference between loving animals and doing right by them. I like to use purebred dogs as an example. The world of show dog breeding is run by dog lovers, not veterinarians, evolutionary biologists, or geneticists. Back in the day, a breeder noticed and liked a slight slope in the hind quarters of German shepherds and wrote breed standards that incentivized the

FPO

FPO

accentuation of that trait—same goes for the cute smooshy faces of breeds like pugs and bulldogs. Today, German shepherds suffer from debilitating arthritis and hip dysplasia before even turning three, and bulldogs have heads so unnaturally big they must be born via c-section then spend the rest of their lives with respiratory issues exacerbated by their janky physicality.

Cats are another good example. In order to get out in front of what is sure to be a global controversy once it's uncovered, I will tell you here and now that I do not like cats. Or rather, I don't understand owning cats. My feelings about *indoor* cats are mostly to do with how bizarre it is to own one as a pet. They are not truly domesticated, and were they even slightly larger, it would no doubt be illegal to possess one, citing the same safety concerns that keep people from owning jaguars. The scratching and biting and home destruction that cat owners are (for some reason) so blasé about might look a little different if house cats weighed sixty-five pounds (the average weight of a golden retriever) instead of eight. As far as I can tell, there's not much difference between owning a house cat and owning a small furry crocodile. But if you love them and they make you happy, then I wish you and your funny little house panther nothing but the best. And I will continue to watch cat videos with the utmost bemusement. I only bring up indoor cats because the fact that people own cats as pets often informs their opinions about outdoor cats. And outdoor cats are a problem—an ecological nightmare with a free pass to spread devastation. It's not their fault. Cats are just being cats. But somewhere along the way, cats have come to represent the most extreme example of our emotional wants getting the better of our environment's rational needs. Few feral or invasive animals in America or elsewhere get as much special treatment from people as outdoor cats do. This sort of preferential treatment of an invasive species, thanks to its link to our beloved pets and despite its status as an environmental hazard, speaks volumes about people's views on animals.

There are two types of outdoor cats; ones that have owners and ones that don't. Both are invasive species and walking global-ecological *cat*astrophes (sorry). Outdoor cats with owners kill birds, small mammals, and reptiles despite being well fed at home—cats without owners are responsible for even more small-animal deaths. Dr. Peter Marra, senior scientist emeritus at the Smithsonian Migratory Bird Center published an entire book about the crisis called *Cat Wars: The Devastating Consequences of a Cuddly Killer*. And he even likes cats.

The numbers are staggering. Before his book came out, Dr. Marra coauthored a report called "The Impact of Free Ranging Domestic Cats on Wildlife of The United States" with his colleagues Scott R. Loss and Tom Will from the Fish and Wildlife Service for the scientific journal *Nature Communications*. Their report calculated that an average of over two billion birds and twelve billion small mammals are killed (just in the United States) by outdoor cats every year and came to this rather chilling conclusion: "Findings suggest that free-ranging cats cause substantially greater wildlife mortality than previously thought and are likely the single greatest source of anthropogenic mortality for US birds and mammals." That's more than communication towers and windows and lighted skyscrapers combined. A 2018 update to Dr. Marra's research, published for the *Society for Conservation Biology*, includes the fact that at least sixty-three species we know of have gone extinct because of outdoor cats. Ornithologist Scott Weidensaul echoed all of Marra's concerns after the release of his 2021 book on bird migration *A World on the Wing*, saying, "The single easiest thing that we could do to save large numbers of birds in North America is keep cats inside."

Dr. Marra also makes clear that in addition to the threat outdoor cats cause to biodiversity, they also "unquestionably threaten humans and other species as hosts of zoonotic disease." I found this particularly interesting considering people's growing concern about animal to human communicable illness. If people can't be bothered to have a tough conversation about outdoor cat management because they don't care enough about other species, then perhaps they might be motivated to have that same conversation when it concerns the health and safety of their own.

You might wonder why I'm ~~ranting~~ talking about cats in this book about hunting. In terms of the difference between loving animals and doing right by them, cats are a purrfect example. It's important to call out dangerous levels of hypocrisy and speciesism within the world of animal lovers and to illustrate the differences between conservationists and ecologists and those who might use the term "animal-welfare activist" to describe themselves. I, too, am for the welfare of animals, but I find myself at odds with these "activists" who prioritize one animal over many entire species. The critically endangered kakapo of New Zealand is the world's only flightless parrot—they're green and adorable and next on the cat-driven extinction list, with fewer than 200 left in the wild. Outdoor cats have already contributed directly to the extinction of nine other species of ground bird in New Zealand. If people insist on steering the trolly car away from the cats, there's no telling how many other species it'll hit instead.

No one is suggesting we make cats go extinct, but we do need to remove the free-roaming outdoor ones. Instead, some folks actively help outdoor cats proliferate, a discouraging example of hypocrisy in how some people relate to animals. If you knew a person who was actively aiding the booming populations of any other invasive species, you'd call them a villain. But that's what "managed cat colonies" are, people feeding and aiding a harmful invasive species. Those who manage such colonies simply see themselves caring for wayward strays, ignoring or losing sight of the results of their actions. It's like an oil spill they've decided to add a little more oil to every day rather than clean up. I like to call them "cat colonialists," but you could also make the case for "subsidized killing factories."

Dr. Weidensaul addressed this issue, saying, "I also have a real problem with so called 'managed cat colonies' or 'trap neuter release colonies.' They essentially become subsidized predator colonies, often near important areas for birds like parks and refuges, places like that. I love cats; I've owned a lot of cats over the years. None of my cats have ever gone outside. They've always seemed perfectly happy watching birds from indoors. I realize that cats can be a really polarizing issue, but this is one thing where cat lovers and bird conservationists ought to be able to find common ground."

The best way to protect ecosystems from cats is to remove the cats. But culling is generally a nonstarter. People who oppose the culling of feral outdoor cats offer no alternative solution other than trap-neuter-release which is wholly ineffective. *Cats do not need their reproductive organs to kill birds and small animals.* If they do, I missed a *very* interesting episode of Planet Earth.

If you want the most uncomplicated version of this issue, look at Australia. The unique evolution of species there has produced some of the most wonderfully bizarre creatures to be found any place on earth (how

FPO

Caption TK

'bout those monotremes?!). I've been in love with the place since I was a kid, because of what a biodiversity wonderland it is–or was. Introduced predators like foxes and cats are leveling Australia's small-animal populations. There are campaigns all over the country to deal with them that include hunting and trapping as well as some fascinating gene-editing and breeding programs meant to teach generations of small mammals how to fear and escape predation (a little sad when you consider the unafraid and even friendly nature of many of the species there). When it comes to issues regarding non-human animals, especially when it concerns their extinction, I'm a belt-and-suspenders man, so I'm pleased to see a multitude of solutions being implemented and on the horizon for the issue–even PETA Australia recognizes euthanasia is on the table. When extinction is a possibility, we have to burn the candle of solutions at both ends.

When lawmakers put plans in motion to cull invasive species from the ecosystems they're invading, they're often met with wildly emotional contempt from people who are either ignorant as to why the situation merits such measures or who just don't care enough about the damage that's being caused to the native ecology. If they did care, they'd have to make a tough decision, because alternative options have proven ineffective. We cannot walk up to feral swine and politely ask them to not eat endangered sea turtle eggs, and we can't hand write letters to cats asking they limit their killing sprees to barn mice and European starlings only. The arguments over how to deal with environmentally hazardous invasive species (including cats) in a way that placates those who are unmoved by the greater ecological issues at stake merely extends periods of inaction, and inaction is what leads to extinction. It all comes back to visibility and perception. The visibility of a neighbor setting out food for stray cats being perceived as good, versus the invisibility of the measurably negative consequences of that action. The visibility of hunters on social media being perceived as bad versus the invisibility of the positive consequences that emanate from their actions. When we allow our emotions to prevent us from seeing the whole picture, everybody loses.

The last twenty years have seen a worrying trend of habitat and wildlife issues being taken out of the hands of experts like biologists, ecologists, and foresters, with emotional motivations overshadowing scientific and environmental economic understanding. It's probably not shocking that a dangerous political schism exists in the realm of wildlife management, as there seems to be one of those divides everywhere these days. The sad part is that both parties want the same things: habitats safe from development and industry and flourishing wildlife populations. But if you love animals, how do you trust people who shoot them to make the best decisions on their behalf? The divide ends up pitting people with an emotional interest against the recommendations of scientific experts (who hunters tend to side with).

Take Spain. In December of 2020, the Spanish government gave in to well-meaning activists who wanted hunting banned in their national parks. Spanish national parks work a little differently than they do in America. The biggest difference, aside from the fact you could hunt in them, is that you can own private land within their borders. Obviously, the landowners who could no longer operate hunting outfits on their property were furious, but beyond that, the immediate reaction from hunters, wildlife scientists, and foresters was a resounding "Yikes. This is not going to end well for the animals."

Before the ban went into effect, the parks were bursting with flora and fauna, each park at the limit of their carrying capacity. After the ban, as you'd expect, wildlife populations started to grow. And grow and grow. As populations rose, habitat declined, becoming over browsed and over rooted, without a chance to recover. The economies that relied on eco-tourism in and near the parks, like hotels, shops, and restaurants, weren't considered at all when the legislation was put in place. Meat from successful deer and boar hunts wasn't being harvested, so anyone who relied on or enjoyed sustainable game meat had to start buying it from elsewhere, just as the cost of living was rising for people in the rural communities.

Just fifteen months after the ban, the explosion in wildlife was already causing headaches for the Spanish government. Animal numbers in the park were now more than double the land's carrying capacity, so the government tried demanding private landowners manage the ever-growing populations. The landowners were curious where they were supposed to get the money to undertake such a job, since they could no longer charge hunters to manage the population for them. A park slightly southwest from the middle of Spain, called Cabañeros, was being so denuded that the government ordered hunters, just in the one park, to kill 5,000 animals in a twenty-eight-day span. If proper management can't be sustained, then estimates suggest 7,000 animals will need to be killed the following year in Cabañeros alone.

Hunters and ecologists following the story the world over were quick to say, "we told you so." And rural communities in Spain began protesting the increasing number of laws being handed down by urban legislatures unfamiliar with the realities of rural life, let alone wildlife management or ecology 101. The animal lovers' heartfelt emotions and best intentions had compromised the park's habitats and made a mass slaughter of animals necessary to save the parks from the animals and the animals from themselves. None of this would be necessary if Spain was entirely uninhabited by humans–Nature has its own system of checks and balances in a human-less world–but when humans shrink habitats enough, Nature's system doesn't work quite as well, and those pockets of "wild" life require consistent regulated management.

Lest we get complacent about US wildlife management's susceptibility to emotion-driven legislation, let's look at a couple examples closer to home. In 2019, California became the first state to outlaw trapping. It was an easy way for lawmakers to appear like they were friends of Nature without making anybody angry or uncomfortable. Unfortunately, it also didn't require them to understand the issue. Trapping, an incredibly time-consuming activity that requires lots of equipment, is very much in decline across the country without government assistance–the California Department of Fish and Wildlife sold 133 trapping licenses across the state in 2018–so the bill made for a cheap publicity win (to actually help the environment, California lawmakers would have to tackle industrial and private water use, the agricultural industry, and their cities' lack of public transportation options, which are all more complicated, connected, and formidable than 133 recreational fur trappers). Unfortunately, California has a nutria problem. These cute (if you like giant rats, like I do) invasive rodents from South America look like mini beavers but with rat tails, and they are devastating aquatic ecosystems across the United States. The California Fish and Wildlife Department needed ten million dollars from the state legislature to aid their nutria-eradication program after they realized populations were growing in 2017. Another bill introduced in 2019 directs even more funds to that project. Guess how they get rid of the nutria? Normally, citizen trappers pay the government for the privilege of trapping nutria, and the money they spend on licenses ends up in state conservation projects. Laws like the one in California shift responsibility for nutria trapping entirely to overworked and understaffed wildlife agencies. This often includes hiring extra hands or contractors to help with the work, which not only means California residents' tax dollars are now paying for trapping to happen but also that those underfunded environmental agencies have to spend more time trapping when they could be using their time and money for other projects and environmental research. In one fell swoop, the anti-trapping taxpayer has become a trapping subsidizer and hamstrung the progress local Fish and Wildlife authorities were making on other, perhaps more pressing, ecological matters. Now we have, 1) less money for conservation, 2) more strain on the underfunded Fish and Wildlife service, 3) critters are still being trapped, and 4) you're the one paying for it.

On the other side of the country, Staten Island is attempting to reduce deer numbers by spending over 6.5 million dollars of tax-payer money (in a program spanning five years) to give deer vasectomies. Wildlife management experts would have preferred a controlled, professional cull, which would have reduced numbers faster and for less money, but government officials agreed the media headache and inevitable ensuing lawsuits wouldn't be worth it. Borough president James Oddo described the plan (of 6.5 million dollars' worth of deer vasectomies, mind you) as "the path of least resistance" in a 2021 report by local news outlet *SI Live*. The plan took longer and cost more than expected, but, so far, it seems to have been moderately successful in keeping fawn births down. However, because wildlife management is an ongoing process, not a one-time fix, the program will have to continue indefinitely. What strikes me as the most unfortunate aspect of this technique is the lost opportunity for sustainable local food. If you read about an island where local game is harvested annually to help supply the community's food banks and school systems, you'd probably think that was pretty cool. But if it's deer on New York City's Staten Island, it suddenly becomes controversial. Not because the deer are any different there than anywhere else, but because the people, so unaccustomed to nature after all this time, have put them in their animal boxes once again, ignoring a chance at living with the wild in favor of tolerating it living next to them.

Words are not what get us into trouble, it's peoples wildly different interpretations of their definitions that do. That plus humanity's preference for black-and-white headlines over what might be a more complicated and nuanced gray area makes for quite the recipe for division and misunderstanding. On my first successful hunt with Fisher, while he was talking to the woman in her suburban backyard and telling her just how much meat one deer would provide him with, I heard her say "Well at least you're eating it." I mentioned it when he met back up with me and said, "What does she think, that we just shoot a deer, cut off its head with a machete and walk out of the woods with it?" I chuckled, but Fisher paused and said, "Actually yeah, I think a lot of people do think something like that."

That day I did shoot a deer, and I asked the butcher if he'd cut off its head for me so I could keep the skull. I took it home and spent a couple days cleaning it so I could preserve it properly and mount it to a small wooden shield. You could call it a trophy. I think of it as a memento which is what a "trophy" is. I didn't start hunting so I could collect trophies. I had collected taxidermy long before I started hunting. And if I was hunting for trophies, then I was failing miserably, as my first archery deer was an antlerless button buck and my first rifle deer barely had four points on his antlers. But I did gain over seventy pounds of venison from the both of them. I went through the meat in under a year, between having it for dinner almost every night and giving a pound away here and there to friends–a necessity at the beginning so I wouldn't be crushed by a meat avalanche every time I opened my freezer door. (Mmmm, meat avalanche.) Even though the meat is gone now, I still have beautiful, tangible, 3D memories hanging on my wall. Do you remember the cheeseburger you ate in 2016? Did it have a "story"? When you see any kind of displayed animal mount–could be a skull, a taxidermy shoulder mount, or maybe the whole critter nose to tail–those are parts (skulls and skin) you don't eat anyway. So why do you care if someone keeps them?

FPO

Caption TK

Maybe it's the motivation you associate with the mounts. You think some asshole took up hunting just so they could put animal heads on their wall to show off. The thing about motivation here is that it doesn't affect the consequences. I hunt as a way to source meat, spend time in the woods and around animals, and support the ecosystems I love. Hunting has come to mean more to me than the environmental and economic benefits it generates, but those benefits are still there regardless. Now let's say some other guy just wants to bring home trophies to show off. We will both have paid into the same system that puts value on habitat and animals, so it doesn't matter what our end-of-day motivations were. It reminds me of the little note in the hotel to "please reuse your towel" because the hotel wants to be "green." The hotel really just wants to save money, and by doing fewer loads of laundry, it does. But it is saving energy too.

What do you think of when you hear *trophy hunting*? Maybe a guy who passes up a small buck so he can shoot a bigger one. Or perhaps a lady who chooses to travel to Florida one spring to hunt turkey instead of in her home state of Vermont. Most likely, your mental picture includes a version of a rich dude who goes to Africa specifically to shoot a lion.

Let's start with the big buck hunter. Here's the scenario, a guy sits in his tree stand and lets three smaller bucks pass him by. He knows there's an older monster out there with a huge twelve-point rack of wide-set antlers, and he's waiting for him. You could call him a trophy hunter, but so what? He still paid all the same fees to Fish and Wildlife, and he's still going to eat it. He just has, arguably, an even more attractive memento if he is successful.

Hunters do take pride in successfully hunting a large, mature animal. Doing so can be a confirmation of hunting ability, not just being in the right place at the right time. Animals don't get to be big and old by being dumb. Take the case of my first deer. As a new hunter, I was more likely to have an opportunity to harvest a young animal like him, since he was as new to being prey as I was at being predator. My buck hadn't had the opportunity to breed yet, but he was in an overpopulated area of the state, and his harvest was part of the management plan there–New Jersey wants fewer deer in some areas than they have now. In places with stable populations, or where you want populations to grow, shooting a button buck would be a big no-no. The folks who are pissed about a big buck hunter's trophy don't understand (or know about) how genetic diversity and government regulations play into a hunter's animal selection.

The argument I hear is that by taking the King of the Forest out of a breeding group, you have altered the selectivity of nature, "unnatural selection" if you will. This is a simple biological theory–take the biggest males out of a population and you take the genetics that made them big with them. But in order for the theory to play out, you'd have to be selecting from an isolated or totally captive group of animals through multiple generations. One instance where scientists did see this occur was in a thirty-year study, published in 2003, of "unrestricted trophy hunting" of rocky mountain bighorn sheep in Alberta, Canada.

Throughout the course of the study, rams in the population in question developed thinner horns and smaller bodies after generations of the biggest males being shot by hunters. The fact that this study took place in Canada actually says a lot about how functional and awesome the American model of sheep hunting regulation is. In the United States, we have extremely restrictive age regulations on what rams can be legally shot. And ram tags are few and far between. This means that by the time a ram becomes legal to shoot, he has already been able to pass his genes along through ample breeding opportunity.

Evolutionary biologist Shane Campbell-Staton studied the effects that the fifteen-year-long civil war in Mozambique (1977–1992) had on wildlife in and around Gorongosa National Park, and he estimates around 90 percent of the large mammal species endemic to the region were killed off by poachers. Elephants were targeted so ivory could be sold to finance the war. Today, the offspring of the surviving elephants are more likely to be born tuskless than they were before the war. Killing of that magnitude was enough to alter the genetic pool of the elephants in the region. Poaching is not legal or regulated–this is why it has its own word. The poacher does not hunt–the poacher poaches. In places with economic instability, where it's not safe for biologists or conservation groups to work let alone have eco-tourists like hunters and photographers, wildlife often pay the greatest price, as they're easily picked off by poachers for the black market and bush meat. Hunters, on the other hand, make up a very small fraction of the mortality rates of game species, and even if every hunter in America was able to pick off the most mature male animal they could find, this would work out to no more than a drop or two in the entire genetic pool. This is precisely why hunters often have age restrictions that favor the shooting of older, mature animals. We want the older, big guys to pass along their genes, then make space for the next generation of younger animals to pass theirs on next. Because it's genetic diversity that keeps a population strong and healthy.

There are many factors that come into play when you're looking at genetics in nature, and legal, well-regulated hunting doesn't represent a

Caption TK

sufficient influence on altering genetic makeup. Or to quote the brilliant and sensual chaotician Dr. Ian Malcolm: "Life finds a way."

Now what about turkey lady? She's hunted turkey in her home state of Vermont, as well as neighboring New York, all her life, and now she's got a new ambition. She'd like to hunt each of the five American sub-species of turkey. She's already got her eastern turkey so now she'll have to make plans to travel to a few states out west and in the Southwest or Mexico to hunt for Merriam's, Gould's, and Rio Grande turkey. Then off to Florida, the only place you can find Osceola turkey. Is she a trophy hunter? She did travel a long way just to shoot an obscure subspecies of bird I'd wager most Floridians don't know much about. But along the way, she injected more money into the conservation funding systems of each of those states than a local hunter did (because she paid more for an out-of-state license). She's still going to eat them. There's definitely a "gotta catch 'em all" aspect to her goals, but, again, what does it matter? When you love something, you want to pursue and appreciate that thing in all manner of ways possible. After the pandemic broke out, lots of home cooks decided to try making sourdough starter. Why? Why do something the hard way? Why take your time if you don't have to? Why make bread from living goop that you must feed and care for, when you can just buy bread that's ready-made from the store? Creating challenges and goals for yourself is fun. Doing stuff the hard way, or the old way, or the traditional way, is rewarding.

Remember Cecil? He was a radio-collared lion being tracked by Oxford's Wildlife Conservation Research Unit in Zimbabwe and was shot in 2015 by a trophy-hunting dentist. There was a huge media dust up afterward because he was a known lion being monitored and he lived in and out of a sanctuary, causing speculation as to whether he was lured off its grounds or not.* I was upset by Cecil's death and heartened by the rush of support for the slain lion in the media. But by the next day, I began to roll my eyes.

---

*Lion hunting in general requires far more restriction than most other species need because of their more complex social structures and breeding behavior. Because of that, it's best if lions are over seven years old if they're going to be hunted. Cecil was thirteen, which is awesome for a lion (their life expectancy in the wild is about 12-16). Cecil's notoriety makes him a useful subject, but he represents a small portion of a larger conversation.

What did the angry posters know about the situation? This was just one more instance of everyone rushing to social media to voice their opinions and listen to their favorite celebrity's take on the matter. And when celebrities, and their limited knowledge of complex environmental issues, rather than scientists steer the ship of conservation, it always leads to a dangerous outcome for animals. Conservation biologist for the University of Oxford and large carnivore specialist Amy Dickman says, "It comes down to the heart of what I find most regrettable about this debate, that the people who have the biggest platforms are often the furthest removed from these topics. And the further I've gotten from the field, the more I've seen the voices about the topics get stronger. And a lack of voice for the people on the ground." Stop what you're doing right now and listen to her interview from 26 March 2021 on *Science + Story*, a podcast with science communicator Bob Lalasz. It's farming (which means habitat conversion and loss) and poaching that will lead to the extinction of your favorite African wildlife, not trophy hunters. If only poachers had Facebook, Instagram, names we knew, or a local news crew to follow them around. Then maybe it would be easier to galvanize public support for anti-poaching measures, which are needed more than anti-hunting ones.

It's pretty easy to look at a photo of a big game hunter nestled up next to a dead giraffe or holding up the body of a leopard in Africa and hate them. The romantic fantasy of undeveloped landscapes in Africa brimming with wildlife sparks stronger feelings in a lot of people than they have for their own local wildlife. People have a personal and visceral response to the animals of Africa (even if they've never been), and I think that's great! I just wish those people would follow through on their passion by following up with the scientists, and that they could muster the same passion for all the wildlife of the world–not just for the leopards of Zambia but for critters like the Australian spotted quoll, and for whatever is living in their neighborhood. Conservation starts in your own backyard.

Properly discussing hunting in Africa would require a whole extra book or two. But it's necessary to bring up because "trophy hunting" is a term that arises in the discussion of hunting in general (within the United States or abroad), and you can't not mention Africa. So here is the shortest way I can break it down. The southern countries of Africa we most associate with trophy hunting do not have unified conservation funding structures like America does, but the economic incentives for protection of land and animals work about the same way. The differences are that most of the money comes from international hunters and conservationists as opposed to local ones, and that Africa must rely on the privatization of their wildlife more than we do in America, where we hold them in public trust. Either way, if you make animals an economic asset, then you want that asset to grow. It would be ignoring reality completely to say there's no corruption in the African trophy hunting systems—there is—the same way there's corruption in any politically or financially charged system. America has poachers too. But the negatives of corruption do not outweigh the positives of the conservation structure.

You know how in real estate they say "location, location, location"? Well, when it comes to wildlife (anywhere) the same is true. "Habitat, habitat, habitat."* The basic cycle of the habitat-wildlife value system goes like this: by putting a price on the head of an animal, you have given that animal value. Not just sunshine-and-rainbows, glory-and-majesty-of-nature value, but actual cash value. And by making the animals valuable, you've just inadvertently made the landscape more valuable as habitat than as farmland for crops or as ranchland for livestock. Because you can't have wild animals without habitat for them to live in. Now that there's money to be made from keeping those animals around so big game hunters can come hunt them, local communities have incentive to care for the landscapes and protect the animals as their own economic assets. Professor Dickman emphasizes habitat often, noting that over 1.4 million square kilometers is set aside for big game hunting in sub-Saharan Africa, and that's more land than comprises the continent's national parks. She points out that if you ask people, "'What's the benefit of national parks?" people would say, 'Well,

*The National Wildlife Federation released a report in 2022 titled *Wildlife Under Pressure: The Dire Need to Combat Habitat Loss*. I was disheartened when I first read the report and saw the statistic that wildlife in America had lost around 6.5 million acres of habitat over the last two decades. My heart sank a little further when I realized they meant per studied species. Mule deer had lost 7.3 million acres, turkey had lost 18.8 million. It also reported that an individual black bear or elk can only ever get about two miles away from "significant development."

you know, it protects land for wildlife.'" Why then, she posits, is hunting so different? If hunting is banned, that land will not become a park, it will become the next location of human development.

One program pioneered by Dickman's team created benefits for communities who had animals show up in camera trap photos. This incentivized villages to provide better habitat for the animals and then protect them. Groups of women would stop younger men from killing lions, as the big cats were now more valuable to the community alive. People often forget that animals don't exist in a vacuum—they must have ample and healthy habitat to survive and multiply, which is increasingly difficult for them to find with escalated human expansion. By making animals valuable, you make the places they live valuable. And because habitat loss (or habitat conversion, as it's called in the conservation space) is the number-one threat to all wildlife species, habitat is just as important as the animals.

Photo safaris in parts of Africa can make good money for local communities and wildlife conservation, but they primarily operate near more heavily developed areas, requiring those places to become more developed to accommodate more travelers. In general, you need way more photo tourists (seventy-seven is the current estimate) to equal the same economic impact of one hunter. The average cost of a one-week luxury photo safari is $5,250 per person. The cost of one elephant license is between $43,000–45,000. And that's just the cost of the license. The hunter still has to pay for a guide and other associated costs of the trip, which can easily tip the scales for that hunt at over $70k. Beyond that specific example, African big game safaris can cost over $200k if the hunter plans to hunt multiple species over a long span of time. While it's true that photo safaris can operate all year, and the hunting season is only about half the year, most tourists will plan their trip around the best weather and highest chances of seeing the most animal activity, essentially limiting their season as well. Hunting safaris offer more economic opportunity and wildlife protections to communities and animals in remote locations. Places that are not built up enough to support large groups of photo tourists are perfectly suited to hunters looking for a wilder experience in the bush, far from the maddening crowds. To be very clear, photo safaris are not at odds with hunting safaris. They both require animals on the landscape to be successful, and both create value for wildlife. Though the value generated from hunting safaris is generally more direct and impactful, while also requiring a smaller footprint (less development), than photo tours, there is no one-size-fits-all model of conservation in Africa. Some places suited for hunting aren't suited for many photographers and vice versa. Also, not every safari is an elephant hunt–in fact, few are. It's quite possible to hunt plains game (like wildebeest and springbok) in Africa for under $10k.

The problem with any tourist economy is that you're toast if the tourists don't show up. Poaching skyrocketed in Africa during the Covid-19 pandemic, when travelers weren't regularly bringing money in. Conservation biologists like Professor Dickman say if we in the West want to keep international wildlife around, we must pay for them, otherwise we don't leave many alternatives for the people who live near those animals.

Hunting safaris make it possible for local communities to profit from the animals that live among them–giving them a vested interest in making sure those animals go to paying customers, not thieving poachers. The money brought in from big game hunters pays for anti-poaching rangers like the ones I wanted to be as a kid. It also affects how locals feel about the animals. In the Western world, we like to fantasize about the almost-mythic animals of Africa, but most of us aren't living with them in our backyard. For many African communities, elephants might as well be giant rabbits destroying their crops. Grazing herds of various species compete with their cattle for grasses. And predators pose a real threat to local people as well as their livestock. The same way coyotes can be seen as pests to ranchers in America, farmers in Africa can see wildlife as competition or a threat to their livelihoods. Local farmers sometimes use the same tactics as poachers, setting snares, poisoning bait, and throwing spears or poison-tipped arrows at "encroaching" wildlife populations to kill or deter them. This problem only gets worse as communities spread further into the bush. (It's not much different than deer flooding the backyards of suburban homeowners in the northeastern United States.) By "selling" some animals to trophy hunters, you make those species and the habitats they live in more economically valuable to the community alive than dead. The system creates a job market that relies on the protection of habitat and wildlife to be successful, rather than a system that requires the removal of both.

This doesn't represent the only human-animal conflict deterrent of course. There are some extraordinary people in the field today who have focused all their work on creating systems of benefit and peaceful coexistence between local communities and wildlife. In 2019, I attended a fundraising dinner for the Hunger Project, another stellar organization that focuses on community training and engagement as the best way to help people pull themselves out of poverty. Rowlands Kaotcha is global vice president for the organization as well as the Southern Africa regional director. This is a story he told that night:

Robert lives outside of the Majete Game Reserve. Maybe you have heard of it? Today, Majete is one of the best game reserves in Malawi, and tourists come from all over the world to see the wildlife. But just fifteen years ago, it was an empty forest with no employment or tourism and only a few antelope living there. For decades, the communities around the reserve were in conflict with the land. They cleared the forest for agriculture and firewood, and they poached the animals both for profit and to protect their fields and homes. Robert Ektera was one of the people who poached. His father started teaching him how to poach elephants, hippos, and zebras when he was six. Six. After years of training, Robert was eventually a master poacher, teaching others how to poach. He believed it was his responsibility to train the next generation, including his own son. Eventually the park ran out of game. The elephants were completely gone, and the hippos no longer soaked in the cool waters. The future looked bleak. So, the government of Malawi partnered with an organization called African Parks to repopulate the game reserve. They knew that to create a sustainable park they needed to work with the communities living outside the reserve, to shift their mindset so they were not in conflict with the land and animals. They asked The Hunger Project to develop a tailored Vision, Commitment, and Action workshop to help the communities see the game reserve as theirs, as a source of prestige, and as a resource for them, not as something to be poached and destroyed. Robert Ektera, the master poacher, decided to come to one of our VCA workshops. During the workshop, we helped him paint a new vision of the future. He saw that poaching was not a sustainable livelihood for him or his son, and that to create a prosperous future for his family he needed to find a way to redirect his energy into something productive. He decided to enroll in our beekeeping program. Through this program, Robert has become a master in natural resource management. With a stake in keeping the land healthy, he has become an advocate for land conservation and preservation. Robert has chosen to use his success as an example and has started to train people in beekeeping. Specifically, he targeted all the people that he trained as poachers, and he motivated them to become beekeepers. When he told me his story in 2017, he said that about 65 percent of the people he has trained as poachers are now part of his beekeeping club. Furthermore, Majete Game Reserve has not had a single rhino or elephant poached since they reintroduced the animals to the park. If it were not for Robert courageously saying to his fellow poachers, "use me as an example of what you can do in your life without poaching," I do not think that would be the current statistic.

In Africa, the expansion of farming and ranching into wildlife habitat is referred to as "the plow and cow." Michael 't Sas-Rolfes is a South African doctoral researcher at the University of Oxford as well as a fellow of the Oxford Martin Programme on the Illegal Wildlife Trade. He's studied conservation economics and the wildlife trade for over twenty years. In a 2016 piece published by the Property and Environment Research Center titled "Better Bred than Dead," he writes, "In Africa, projected forecasts of human population and economic growth—coupled with needs of food security—suggest that the pressure on wildlife will soon intensify. It is also instructive to look at examples of the addax, dama gazelle, and scimitar oryx—antelope species known collectively as the 'three amigos.' These species were mostly exterminated from their home ranges in North Africa for food by hungry locals during times of civic unrest."

When people think about protecting African wildlife, they usually stop right there, at the wildlife. That's something I've been guilty of too. I didn't give a rat's ass about some dude's crops or cows, regardless of whether he was in Malawi or Montana. "How dare you compare the importance of some farmed cattle to African lions or mountain lions?" I thought. But the reality is we only get to save the lions with the help of those cattle-farming locals. And sadly, it's these communities of local people who are often forgotten when Westerners impose their understanding of conservation on Africans. Jonathan Adams and Thomas McShaWite in *The Myth of Wild Africa*, "Conservation cannot be done 'to' or even 'for' or 'with' Africans. Conservation must be done by Africans." Personally, I'd rather see local communities build their economies through eco-tourism and wildlife-benefit incentives, which includes hunting, so the land becomes more valuable as habitat than as farmland. It's the same way I'd rather see forested woods than a corn field, or bison roaming the prairie instead of cattle. If there's a way for people, habitats, and wild animals to all thrive at once, why wouldn't that be our first choice?

Travis de Villiers, a South African I know from the taxidermy world, was interviewed on a science and conservation podcast called *Specimens* and reminded listeners that locals and poachers kill indiscriminately. Meaning there's no consideration of age or sex or population stability among species, the way there is in areas with regulated hunting. Having a hunting economy (for locals as well as foreigners) makes all the animals within that economy an asset, so the decisions of which animals can be removed without effecting the population are left to professional guides and wardens who do nothing but study that very question. After all, the guides wouldn't have much of a job if the animals they were guiding people to disappeared. Farmers, on the other hand, are generally thrilled if they don't have to worry about wild animals eating their crops or killing their livestock. Enabling local communities to profit off animals, provides personal incentive for the animal's protection and proliferation. Travis likened well-meaning but misguided anti-hunters to a new wave of Western colonialism–people dictating what they think is best from far away, on their smartphones and in the comfort of their own air-conditioned homes. "We know what we are doing with our wildlife," he says.

Other hero problem-solvers in this field include zoologist Dr. Lucy King, head of the Human-Elephant Co-Existence Program for Save the Elephants, who helped to develop a new system of beehive "fences" that deter elephants–elephants can't stand bees–while providing the added benefits of busy pollinators for local crops and "elephant-safe" honey for the community to use and sell; conservation biologist Dr. Leela Hazzah, cofounder of Lion Guardians, a badass program that trains and employs Maasai warriors to become the defenders and monitors of the lions in their region rather than their adversaries; and Dr. Moreangels Mbizah, who studied Cecil the lion for three years and is the founder of Wildlife Conservation Action. Dr. Mbizah gave a brief TED talk in 2019 about the importance of local engagement for the benefit of habitat conservation and wildlife protections and describes WCA's approach to their work by hammering home the importance of community engagement. "Although one of our ultimate goals is to conserve wildlife, we also focus on empowering communities living adjacent to wildlife areas and improving their livelihoods. We believe that seeking economic empowerment opportunities will have a net effect of reducing the community's dependence on illegal activities i.e., bush meat poaching and illegal wildlife trade." A similar model powers the organization Blue Ventures, but with a focus on teaching sustainable wild seafood management for coastal communities as opposed to terrestrial. They helped implement octopus fishing closures in Madagascar for six months, then asked people to only use certain fishing equipment and only take a certain number of octopus at a time (sounds like hunting seasons, bag limits, and methods of take to me). The restrictions led to bigger octopus and more of them. It's groups like these that create the next generation of local conservationists who can source from the wild without depleting it.

Costa Rica (while far from a perfect analogue to Africa) has proven to be a success story in the realm of "structural transformation" as economists would put it, and part of that shift is the cycle of sustainable farming, wildlife viewing, and the safety the country affords travelers, making it an attractive location for tourists. My friend and I took a horseback ride through the mountains there once, and I asked our guide, George, if he ever got sick of all the tourists. He laughed at me like I was a moron and said, "No. I used to work all day in the sun on a sugar cane farm, breaking my back. Now I make more money taking people on horseback rides through the trees and showing them the animals." If people pay more to see the jungle, then we might end up with more jungles and fewer sugarcane farms. We saw a tamandua (a cute little arboreal anteater) on that ride, and George was as giddy as we were about it.

Anti-hunters and sustainable-use detractors will often hear the "hunting is conservation" line and furiously query why someone wouldn't just give that money to conservation directly. It's true, people could do that, but they don't. People pay for experiences. Hunting, at home or abroad, is an experience. So are photo safaris–every tourist wildlife photographer is someone who could have just given their money to conservation too. Even public radio listeners are more likely to donate when they know they'll get a tote bag out of it.

The biggest problem in what is a confusing ethical miasma for most people is the public perception of the hunter as the villain (because they are the most visible) instead of habitat loss, human expansion, and the illegal-trade black markets (those in China, Laos, Myanmar, and Viet Nam being the main culprits) that put so much value on rhino horn and ivory, bear gall bladders, shark's fins, tiger penis, and pangolin scales, just to name a few of the hits. Most of these parts played a role in what once was traditional "medicine," but the market is fueled now by Asia's rapidly growing wealth and a desire to display that wealth. Nobody needs rhino horn to get boners now that we have pills (that actually work), but the ownership of rhino horn can be seen by some as more impressive than any house or car.

Not only is buying horn and ivory seen as a status symbol, but–more disgustingly–they're viewed as an investment. Imagine buying up as many Picassos as you can get your hands on. Then shooting Pablo in his studio. Now you've got a little stockpile of his art to sell, and he's sure as hell not going to make any more. What's extra sad is that rhino horns can be removed without hurting any rhinos. The horn is made of keratin like your fingernails and can be sawed off with relative ease and no harm done to the chubby unicorns. Horn shaving is a common practice among conservationists trying to make rhinos less appealing to poachers. Because poachers do not use tranquilizer darts.

If you want to take the sticky-wicket economic debates farther, check out John Hume the rhino farmer. Basically, he wants to create legal rhino farms across southern Africa that could be run by local communities. The locals would copy his methods and own rhinos the same way farmers own sheep (as delightful as the image of rhinos bunched together in flocks and tended to by a crook-wielding shepherd is, it's not reality–they actually roam independently on large portions of land). So instead of shearing wool, they saw off a horn every couple of years. Now the rhinos are happy because they aren't dead, the farmers are happy because they're making money (off a native species no less), and the idiot who thinks rhino horn will return virility to his sad, shriveled pecker is happy. When a rhino reaches full maturity, past its breeding prime, a hunter can pay to stalk it. Mr. Hume and those would-be local rhino owners have to be able to legally sell their rhino horns, though. That means making the trade in horn legal. If it's illegal, then he can't sell it, and if he can't sell it, then he can't make money, and if he doesn't have enough money, then he can't afford the blockbuster spy-movie level security protections his rhinos (and all rhinos) need to be kept safe from poachers.

I spoke to economist Peter Reuter about Mr. Hume's dilemma, and he was supportive of how the system would "take advantage of rich human predators." Reuter wrote the book(s) on the economics of crime, especially as they relate to drugs, which was precisely why I wanted to talk to him. I told him there was a 30 percent rise in the rhino population in Namibia after the legalization of rhino hunting there, and he barely needed a second to respond with how "poachers would become a threat to the legal market–that's good–it creates incentive to stop them." I told him wildlife rangers occasionally become poachers when they can profit off animals in the black-market trade. He was familiar, "Gamekeeper turned poacher is no new phenomenon, but the gamekeeper is less likely to become a poacher if he's a stakeholder with the other locals." I love talking to economists. No emotional whinging, just numbers, logic, and the simple predictability of human behavior.

The International Union for Conservation of Nature (the IUCN) wrote, "The costs of keeping rhinos safe have risen greatly and live sale prices have significantly decreased over the last decade, reducing incentives for private landowners and communities to keep rhinos. With around half of white rhinos and close to 40 percent of black rhinos now conserved on privately or community managed land, the trend toward rhinos being increasingly viewed as costly liabilities could threaten to limit or reverse the future expansion of the species' range and numbers." What this translates to is that "trophy bans" and increasing restrictions on hunting and legal wildlife trade mean there's been less money and incentive for people to bother breeding or protecting rhinos.

If we could stop an entire culture from wanting rhino horn and elephant ivory and pangolin scales, then there wouldn't be a black market for them at all, of course. The illegal trafficking in wild animals is almost equal to the drug trade in both scope and scale. In fact, rhino horn goes for more than cocaine and three times the price of gold by weight. In 2017, big game hunters legally killed 1,100 elephants. But poachers topped the charts that year, killing 30,000. Unfortunately, there's no "Illegal Wild Animal Trafficking Professionals" Facebook group, and "Barrett the Elephant Poacher" doesn't have a Twitter feed. We could leave some awesome death threats there. Guess we'll have to settle for that one guy and his legally taken lion.

You don't have to like the idea of big game trophy hunting in Africa, and you don't have to like the people who do it. But you should acknowledge that hunters benefit from there being more animals on the landscape, not fewer, and that they pay into an economic cycle that makes the animals and the land valuable to keep around, just like other eco-tourists. Poachers benefit from there being fewer animals—it makes their products worth more. So, if you want to hate someone, make it the guy who buys rhino horn to cure his cancer (that's what the black-market ad execs came up with after Viagra proved to be a more reliable boner source). Though I would love to hunt in Africa one day, the same way I'd like to hunt in New Zealand or Alaska, if all big game hunting in Africa ended tomorrow, I wouldn't be personally upset that my chances of shooting one of the Big Five was now gone. This isn't personal. What I want most is to have these wild animals around forever, and I'll support any system or combination of systems that are proven to promote their future existence on this earth. Like it or not, hunting in Africa is a functional method of generating revenue and an incentive for local engagement and the conservation of species and habitat—the same way hunting accomplishes those things in the States and has brought species back from the brink of extinction while doing it. If we saw greater populations of animals with lower instances of poaching in areas that only allow photo safaris, that would be great, but, so far, the research does not reflect that in wildlife populations. Nor will we be able to implement a one-size-fits-all solution for every country and region of the continent. We must ask ourselves what is working best, not what we like the most. We must listen to the science, not the celebrities. If, tomorrow, we find out the best scientifically recommended way to protect species, empower local communities, preserve habitat, and stop poachers is actually nude mud wrestling, then I'll get my ass waxed and be on the next flight to Zimbabwe. In the meantime, if we stop hunting in Africa today, we will strip animals of their land, aid poachers, and rob local communities of an environmentally beneficial and sustainable economy.

The biggest issues facing southern African wildlife are: 1) habitat loss, 2) poaching and the social acceptability of using illegal wild animal parts, and 3) local community engagement. If the people who live near the animals want them protected, either because they love them or they profit off them, then regulations, social norms, and laws will be implemented to reflect that. Big game hunters, "trophy hunters" (as gross as they might be to you), are not a threat, they just make easy targets.

10

Hunter Blatherer

Debates within the Hunting Community

Cats, trophies, and hunting in Africa are broad issues of local and international discussion—emotional human issues, if you will, that affect people around the globe. Some topics are more specific to the United States and can stir discourse among hunters and non-hunters alike. As I was able to look at these matters fresh, with my critical conservationist's eye rather than my emotional, animal-loving heart, my understanding shifted. I still describe myself as an animal lover but acknowledge that that's the least useful thing I can do for animals and their habitats.* Individuals are important—so are groups. But it's impossible to value the individual and the group with the same weight at the same time, which is why many hunting-related topics may forever be resigned to the unsatisfying purgatory of "it's complicated."

I was once up in the Catskill mountains visiting friends and attending a fundraiser for a conservation group I'm a member of. One of our friends, a local farmer and forager, led a Foraging 101 walk that wove us through woods and fields in search of easily identifiable wild greens. After the walk, I was chatting with the forager and two friends who are hunters and something about trapping came up, which led to her telling us how on the farm they use Havahart live-capture cages to trap woodchucks (also known as groundhogs), then they drown them in the pond. Our faces went white as sheets and our jaws dropped open. Seeing the looks of horror plastered on our faces, she asked with unfazed curiosity, "Why? What would you do?" "Shoot them!" We all exclaimed in unison. She explained they don't keep guns on the farm. My friend chuckled as he said, "They must be pacifists." One person's humane is another person's horror.

My personal mores in regard to killing animals generally boil down to minimizing stress for the individual, so, before I knew much about it, I'd always had trouble stomaching the idea of trapping. But having learned my lesson with hunting–that it's best to learn about a thing before you declare yourself against it–I eventually decided to get my trapper's license.

Before I even opened the student manual and worksheet, I was sure that, if I ever did go trapping, I was only going to use body-gripping traps for semiaquatic species, like muskrat, nutria, and beaver. Similar to a big mouse trap, the force body-gripping traps generate when snapping shut on the back of a small animal's head is enough to dispatch them quickly, and their submerged, aquatic location practically eliminates the chance of catching something other than your intended quarry. My *New York State Sportsman Education Student Manual for Trapping Furbearers* recognizes the emotional gravity of trapping and the mitigation efforts someone like me might take, plainly stating that "just as trapping is not for everyone, land trapping may not be for every trapper."

My trapping class, like Hunter's Ed, was fun. I rarely get the opportunity to nerd out with a group of people so interested in mustelids and semiaquatic rodents, and I was surprised but pleased to see it was full. Maybe that was because our class was really the only game in town–one dude drove eight hours to be there. With fur prices so low, most of the people in

---

*Emotion in the wildlife sciences isn't all bad. What we have to remember is that conservationists and biologists got into conservation and biology because they love animals and nature. Arguably their emotional investments are greater than anyone else's because they decided to spend their entire careers studying and aiding wildlife. What rubs the scientific community the wrong way is when the group with the emotional claim gets more attention than the group with the emotional and scientific one. If you love animals, why wouldn't you always choose the scientifically recommended best practices to ensure their survival? It drives scientists nuts.

my class were more interested in managing small-predator populations on their property than in turning a profit. One guy I chatted with during our lunch break was there for the same reason I was, he hunted but felt unsure about the ethics of trapping, so he decided to take the class just to check it out and learn. We both agreed it was too much work and equipment to take up seriously, but that we might try to trap some muskrat. There was also a whole family there, a mom, dad, and son. They thought trapping might be a good family activity because the dad liked to be outside and wanted to encourage the same love of the outdoors in his son (and keep him away from screens), and the mom had some tailoring skills and wanted to start making handmade fur hats to sell.

Trapping class was more focused on humane practices than I was expecting. When you hear *trap* do you think of large, sharp, interlocking cartoon shark teeth snapping shut on an unlucky foot? Those types of traps are long gone and illegal in the United States, replaced by smooth metal and rubberized clamps, often with a space between the sides so they can't fully close. Should the wrong critter find themselves caught in one, they can be released without harm. The situation is stressful (because no animal enjoys being trapped, in a cage or on a trap line), but foothold traps remain the preferred trapping mechanism for many biologists involved in wolf relocation and repopulation programs. Body-gripping traps are used for smaller species. I was surprised to find that my instructors were more critical of body-gripping traps, which are designed to kill, than of foothold traps, which do just what their name says–hold. If you catch something in a foothold trap you didn't mean to, you can just let it go. Foothold traps also aren't as harmful to animals as I'd been led to believe–any injury is more likely to be a chipped tooth from biting the trap than anything to do with the captive's foot. I didn't know anything about swivels, but in class I learned you need at least three per chain (connecting the anchor to the trap), and the more the better, because the swivels keep the chain from getting tangled and becoming shorter if the animal moves around. There was also a full review of the Best Management Practices for Trapping–the ideal "actions, equipment, and techniques that improve animal welfare and avoid the unintended capture of other animals." This program was developed in the mid-1990s by the Association of Fish and Wildlife Agencies to document improvements in trapping technologies and improve the welfare of captured animals. Developing it was an impressive project that combined the work of biologists, trappers, veterinarians, and scientists from the Universities of Georgia and Wyoming. The information disseminated to students is updated as new techniques and equipment are developed.

FPO

The placement of traps is another aspect of trapping I hadn't considered–it's calculation not coincidence. Traps of any kind are placed strategically to target the specific species a trapper is after and to minimize the chance that any "non-target" species might get caught. Accidents like a dog getting their foot stuck in a trap can happen, but they're rare. New Mexico outlawed trapping on public lands in 2021 because a dog was caught and killed in an illegally set snare, an utter tragedy I'd never recover from if I were I the dog owner. But the state's reaction was overboard. Snares were illegal in the place the dog was killed precisely because the area received that sort of traffic–the person who set the snare was a poacher not a trapper. Drunk driving is illegal, too, and dangerous to everyone else on the road (and sometimes not on the road), but we don't outlaw cars every time someone does it.

Trapping is too valuable a tool for the management of overpopulated and invasive species for it to be outlawed. Maryland's fragile Chesapeake Bay ecosystem made a fantastic recovery from the devastation caused by nutria, and it wouldn't have been possible without the implementation of their Nutria Eradication Project, the traps it relied upon so heavily, and the aid of citizen trappers who effectively allowed wildlife management to outsource parts of the job while paying the government for the privilege of doing it.

Trappers know that trapping can be a sensitive subject. The National Trapper's Association understands the uphill battle of public perception more than most. A sentence in their guidebook predicts the outcome of careless and illegal practices, noting "If you deliberately break the law . . . you bring reproach upon all trappers." As long as trapping continues to be a successful tool for wildlife and habitat management, and the sale of trapping licenses continues to fund state conservation projects, I'll continue to wrestle with the occasional discomfort aspects of it cause me.

Meat is an easier sell for most people than fur. Food is a necessity. Fur seems like the opposite, an unnecessary luxury fashion choice. When thinking about fur, I had to remind myself of those photos I used to keep on me while working as an outdoor gear outfitter—the fluffy white sheep, grazing undisturbed in their New Zealand paradise versus the petrochemical factory, a metal monolith churning out plastic from oil and spewing thick billows of smoke and noxious gasses into the air. The same way a hunter's table fare is environmentally superior to store-bought factory farm steak or a monocrop-based processed patty, so is a trapper's fur hat superior to a synthetic beanie made of plastic-derived acrylic, polyester, or Lycra.

I know that *FUR* can raise some people's hackles, but for the hunter or trapper, it's an animal product that has a use just like meat, and it creates incentive for the management of small predators like coyotes, raccoons, and weasels, and of rodents like nutria and beavers—raccoon and beaver are commonly eaten as well. If I made myself a fur hat from trapping, it would be the equivalent of the meat in my freezer from hunting, a sustainable and environmentally considerate alternative to a synthetic. And the money would go back to Fish and Wildlife, not a clothing manufacturer. (When I realized this, I felt silly for just figuring it out. Oh, you mean the way we did things before rampant industrialization blighted the land with factories was better for the Earth? Duh, past-me, duh.) The market for nutria pelts, for example, has been bolstered by its reputation as "guilt free fur"—which is great. Without a market for nutria fur (or a government bounty), trappers wouldn't have the incentive to trap for them at all beyond their own interest in making a swamp-rat hat for themselves or a friend. And, as we've discussed, if the trappers don't (or can't) trap, then it's up to the taxpayer to pay for state-funded wildlife management trapping programs, whether they like it or not. So go ahead and luxuriate in a fine nutria furkini or coat, knowing it's the byproduct of sound ecological stewardship.*

When I was a kid, I had a raccoon hat that I loved—the classic pill-box style with the striped raccoon tail hanging down the back. I wore it to play out the Daniel Boon and Davy Crocket adventures I read about and always had it

---

*A *New Yorker* article from March 2022 by Rebecca Mead titled "Should Leopards Be Paid for their Spots?" examines the idea of a species royalty for animal prints and symbolism. Can you imagine if every designer wanting to use leopard, tiger, zebra, or giraffe print had to pay a conservation royalty? I'm all for it. More wildlife excise taxes please.

on around the house. I lived in southern Florida until the sixth grade, so it was too hot to ever wear outside, otherwise I'm sure it wouldn't have left my head. One evening, my parents got me a new babysitter, and, in an effort to educate her about my likes and dislikes so she could more effectively babysit me, I told her that I loved animals (and that I liked to put gobs of peanut butter into vanilla ice cream) and went on to show her all my favorite animal books that she could choose from to read (not to me, mind you, just for her to read if she wanted to learn about pinnipeds), as well as my extensive collection of Schleich, Bullyland, and the now-discontinued Britians brand animal figurines. (British paleontologist Darren Naish wrote an article for *Scientific American* about "zoologically minded people's" affinity for these toys. I still have all mine.) Then I proudly showed off my raccoon hat.

She was instantly dismayed and explained that the purchase of my hat would lead to more raccoons being killed so a new hat could be made to replace the one I had bought. It was a startling lesson in economics–I stopped wearing my beloved hat.

I never forgot that moment, but it raised questions for me as I got older. Did my hat's raccoon come from a raccoon farm or was it trapped wild? And if buying something led to the production of a replacement, then why on earth were "animal welfare" groups tossing paint on people's coats? Those people who claimed to care about animals had just single handedly led to the creation of a whole new coat!

Left alone, fur can last lifetimes. My friend Ryan, a lawyer, machine learning expert, and all-around goofball, bought a vintage raccoon fur coat not long ago that, upon inspection of the inside pocket label, was discovered to have been handmade in Canada in the late 1930s. How many of us are walking around in seventy-plus-year-old coats? The longevity of natural materials like fur and wool, along with how they are created (by nature, not a factory) makes them far "greener" than any synthetic could ever claim. So, I had to think about which was more important, the life of the raccoon or the life of the earth it lives on.

Just like my slow "come to Nature" awakening about hunting for my own meat, I started reconsidering my relationship with fur that came from citizen trappers. I don't expect to be dressed head to toe in pelts any time soon, but could I trap some muskrats or a beaver and make my own hat?

Maybe a pair of cold-weather gauntlet mittens? Something that will last forever, come from a renewable resource, and fund conservation and habitat restoration all at the same time sounds a lot better to me than some cheap synthetic from a fast-fashion brand.

A dog is perhaps the best tool and companion a hunter (and obviously anyone) can have, and there is no happier dog than the one working hard next to their owner. That could be a border collie trained to herd sheep or chase birds off an airport runway, a drug- or truffle- or you-name-it-sniffing beagle, or a Labrador retrieving ducks from a frigid lake or beers from the fridge. I've never had any ethical debates with myself about the use of dogs in the field to point out or retrieve birds. Quite the opposite—I daydreamed about walking through the woods and open grasslands with my shotgun and trusty dog at my side long before I ever considered taking up hunting. Any time a hunting dog is in the field it's the best day of their life. They're doing the thing they were literally created to do. Working together and having an activity always seemed more fun to me than just walking a dog around for an extended potty break—back to my issues with hiking I suppose. It was the practice of using dogs to "tree" animals that made me squeamish.

Treeing is simple—the dogs pick up the scent of a black bear or mountain lion or raccoon and, if they can find it, chase that animal up a tree. Then the hunter shoots their quarry from the ground. Some hunters don't consider this "fair chase." One of the things that gives animals an advantage over hunters is the animal's awesome sense of smell and the hunter's shitty one. By adding two to eight dog noses to a chase, you've just made evasion more difficult for the animal you're pursuing, evening the odds or shifting them in your favor (hunting with dogs is easier than hunting without them, but that doesn't mean it's "easy" overall—using dogs does not guarantee success, but it does increase your chances). However, my personal concern came not from the levelness of the playing field but from the prolonged stress of the animal being pursued and cornered.

Folks who "run dogs" have since explained to me that, while being hunted by dogs might be stressful for an animal in the moment, it increases positive selectivity by the hunter. Nobody wants to cause an animal undue stress, but by seeing that animal close up in a tree, the hunter can now, for example, make sure the bear is a boar (male), because they can get a better look at him. If it's not, then the hunter can just walk away. I had never considered that, because I had never thought about it long enough. What's more, if animals like bears and mountain lions are kept skittish to the sounds of barking dogs and human voices, that's a good thing–it keeps dogs, people, and mountain lions safe, especially as people continue to move into their habitats.

When I was little, my dad and I fished for trout together and ate them; I understood that. (I mean, sometimes we ate them. Sometimes black bears would steal them off our grill to my great delight.) But on one fishing trip together, after catching a fish I wanted to eat, my dad told me we were going to let him go. I was stupefied over my father's sudden magnanimity, and then baffled–if we weren't going to eat the fish, then what the hell we were doing on the river?

It was hard for me to grasp that my dad would catch a fish just because it was fun. Today, I can admit angling is fun, and I love having the excuse to stand in a river, but my feelings about catch and release haven't changed much since that day. I'd rather catch two fish and eat them than slightly risk the lives of ten or so by fighting them and throwing them back (there's always a slight chance the angler's fight with a fish could lead to the fish's death after it's been released). Despite my preference, however, the recreational angling industry is worth losing a potential small percentage of fish. The same way there is a money-generating economy built around the conservation of land and animals for hunting, there is a similar economy built around the angler, and that economy only generates money if lakes and rivers are clean, healthy, and filled with lots of fish. The plant and animal life surrounding a river and the water that runs through it will dictate if the fish will be healthy, but it's often people and their money that lead to the fish being safe. Not from osprey, but from farming, mining, drilling, and development.

If anglers and canoers and other outdoor recreation enthusiasts weren't out enjoying Minnesota's Boundary Waters, who would fight to protect them from copper mining interests? What about smaller mountain rivers and streams—the ones too small for a kayak? The same way the elk hunter wants to protect land for elk, the fly-fishing fanatic wants to protect even the smallest stream for little greenback cutthroat trout, to name one example. Recovery efforts for this fish were taken up by Colorado Parks and Wildlife, an agency funded with the money generated from fishing and hunting licenses. The trout went from presumed extinct in the late 1930s, to rediscovered and listed as endangered, to recovered enough to be delisted and declared Colorado's state fish in 1994. Recovery efforts and habitat conservation programs continue for them today, usually with the aid of specialized groups like Trout Unlimited.

The Congressional Sportsman's Foundation summarizes Sunday hunting bans thusly: "Sunday hunting bans are one of the last remaining examples of the puritanical blue laws that were initially designed to encourage church attendance... other activities that were illegal on Sundays included: opening a store for business, drinking alcoholic beverages, and tilling your fields."

There are still states with arcane, unconstitutional, Sunday hunting restrictions, and they all play out a little differently. If you're planning on hunting in one of these states, do not use the following list to make your plans. Check the state's website for the most up-to-date and comprehensive regulations. After the uptick in hunter numbers nationally during the Covid-19 pandemic, a handful of the last states with Sunday hunting bans, recognizing they were losing both conservation dollars and the injections of cash out-of-state hunters bring to rural areas, started to repeal the laws. Hopefully, the trend will continue as increasing hunter numbers make more states realize what they're missing. Currently, Maine and Massachusetts are the only two states with total outright bans on Sunday hunting. Maine is "considering" a repeal, as are most of the following states, as repeals gain popularity. This list is just an illustration of the wackiest prohibitions that still linger in state laws:

Virginia: repealed the last of its Sunday hunting restrictions in April of 2022.

Pennsylvania: no Sunday hunting except for crows, coyote, fox, and some case-by-case exceptions. Up for total repeal in 2022.

New Jersey: no Sunday hunting except for stocked game on commercial property and bow hunting for deer on state wildlife management areas and private property. Sunday mornings during the season, you can hunt possums and raccoons between midnight and one hour before the legal sunrise. (Okay, New Jersey, calm down.)

Maryland: it's legal to hunt "on certain Sundays in certain areas" (it actually said that on the Maryland Guide to Hunting and Trapping website), but you can practice falconry for seasonal game birds on Sundays if you have a permit (naturally).

Delaware: it's illegal to hunt on Sundays with the exception of deer on private land as well as some public lands. But only 20 percent of Delaware's land is public. So...

Connecticut: some private shooting preserves allow Sunday hunting if they're licensed, as well as some dog training and field-trial events as long as permission has been granted from the town. Wild turkey can never be hunted on Sundays, though there are some Sunday hunting exceptions for deer during archery season.

These laws are problematic beyond the intermingling of church influences with government. Many Americans don't have the luxury of hunting during the week. This both limits the access of locals and robs the state of potential conservation and economic funding that would otherwise be brought in by traveling out-of-state hunters (the Theodore Roosevelt Conservation Partnership estimates that deer hunting alone generates forty billion dollars in annual economic activity in the United States). Unlike outdated unenforced state laws like where it may or may not be illegal to keep an ice cream cone in your back pocket Sunday hunting laws create a real loss for conservation and local economies.

A 2019 survey conducted for the Association of Fish and Wildlife Agencies shows that 87 percent of Americans approve of legal, regulated hunting for meat. It's when you start talking about specific species that approval ratings start to jump around. One graph from the same study shows approval of turkey and deer hunting at 78 percent, the hunting of wolves at 39 percent, and mountain lions at 38 percent. Although many hunters have different opinions on what's appropriate when it comes to hunting predators, nothing matches the white-hot distain for predator hunting I have seen from anti-hunters. In the United States, all species are regulated in the same way—to sustain (if not encourage) population numbers and health—but most people view predators as a special case.

Predators just do that to us; I won't pretend they don't. They have forward-facing eyes like we do, so it's easier to look them in the face, and, from the day we get our first Teddy bear, we are groomed to see predators as special—there's fewer of them in relation to prey species, and they're more generally regarded as magnificent or cute. I'm not immune to the charisma of these creatures; bears are my favorite animal, and while I don't disapprove of bear hunting, I'm still not sure if I could shoot one myself. I don't think I'd be sure unless I was bear hunting. And even then, I don't know.

It's usually the "you don't eat them" argument that suspicious people use while decrying the hunting of predator species. But both bears and mountain lions are regarded as good eating by hunters. Mountain lion, though I've yet to have the pleasure of trying any myself, is often described as similar to pork. And bear actually yields more product to the hunter than most game species, because, in addition to the meat, a bear's fat and hide are so highly regarded. Some people might consider meat to be the *best* use of a hunted species, but that doesn't mean the other uses are worthless. Predators, like coyotes, wolves, and bobcats are hunted or trapped for population control, fur, or human-animal conflict rather than meat. We manage deer and we eat them. We manage raccoons and coyote, and we use their hides. As we've discussed, if hunters didn't pay do that, then taxpayers would.

If hunters want predator hunting to be less controversial (and therefore more likely to stick around), then they have a lot of work to do. The resentment ranchers and hunters can show toward wolves, coyotes, and mountain lions for the killing of livestock and wild game damages public perception, making hunting predators look more like vanquishing an enemy than a normal part of wildlife management. A gardener might regard a rabbit nibbling their greens as a pest, but many ranchers and elk hunters hate predators like wolves, coyote, and mountain lions. And it's hard to blame them when that predator is found nibbling on their sheep. Like it or not, if they're hunting legally, it shouldn't matter if the hunter or rancher who shoots a coyote hates that coyote–the outcome is the same–but in order for public perception to accept this, it's crucial to spread awareness of how wildlife management systems function in America. This would hopefully help people see the shooting of one mountain lion as not so different from the shooting of one deer, in terms of population strength and longevity. There are more deer than mountain lions, and that's why there are fewer mountain lion tags allotted each year than deer tags.

In California, a moratorium on hunting mountain lions had been in place since 1972, which helped the population bounce back after bounties on the animals had depleted their numbers dramatically in the early 1900s. In 1990, California outlawed mountain lion hunting completely. But each year, Fish and Wildlife professionals still have to kill "problem cats" who snatch people's pets, and ranchers can still obtain depredation permits to kill a mountain lion after attacks on livestock. With the California mountain lion population now stable, the state could open a limited-draw hunting license lottery that would generate funding for state conservation programs, but the optics of "let's start hunting mountain lions again" are far too inflammatory. So, the cats will continue to be killed but without generating funds for California wildlife conservation.

Perhaps more so than any other predator, wolves occupy a weird space in human-animal relations. First came fairytale vilification, then widespread eradication, and now mystical fascination. I get it, they're social, playful, powerful, cute and fluffy, and they look like our doggies. But all that history and lore heaped on a species can spell trouble. The human tendency to mythologize wolves is what got them into hot water in the late 1800s–when they were effectively made extinct in the continental United States–this time around, it's the fact people think they're so irreproachable and magical that's becoming a problem for them.

Predators have their place in the ecosystem, and I want there to be plenty of healthy predator populations in the wild, but that does mean managing them just like every other species. There's a long sad history of how wolves, mountain lions, and grizzly were eradicated from much of this country in the past, but, unfortunately, we can't look back to historic ranges and populations as our guide for modern day goals–there's just too many people and not enough land. Having wolves in Alaska is not the same thing as having wolves in Colorado.

Wolf reintroductions are always a hot topic. A friend of mine who grew up in Denver and moved back in 2017 after living in NYC, told me she voted for wolf reintroductions there. I was aghast, not at how she'd voted, but at how she'd had the opportunity to. My friend has many talents and interests–she works in tech, likes to fix up old motorcycles with her grandfather, snowboards, and mountain bikes–she does not, however, have a background in ecology, natural resource economics, or wildlife management. (All types of folks make up the wolf reintroduction fan club, but nobody wants wolves back more than car insurance lobbyists, who note around 25 percent declines in motor vehicle collisions with deer in areas with wolf populations.) As much as I love wolves, not every landscape is ready for their reintroduction. And if human populations continue to grow and spread, they may never be. Bringing wolves back to areas of their historic range might turn out to be a celebrated environmental come-back story, but if it is, we can't lose our minds when wolves reach a number subject to management–too many animal lovers view scientifically accepted management techniques as if they are a new wave of eradication methods. When hunting bans are lifted, that's a good thing–it means the population has reached a number that hunting can sustainably manage, just like the elk reintroductions in Kentucky and Virginia. Though it may seem counterintuitive, legal hunting of predators like wolves and mountain lions means the animals are doing well.

FPO

Trapping and the use of dogs are hotly contested topics within the American hunting community. They can be as controversial between groups of hunters as between hunters and anti-hunters, with hunters who don't trap or use dogs showing apathy toward the regulation or outlawing of those practices, while the group who does trap or use dogs points a wagging finger and says, "Mark my words, they'll come for your deer season next." The fear is that one practice being outlawed will lead to all practices eventually being outlawed. This sets in motion a conversation about slippery slopes and shifting baselines that is less about those specific practices and more about the best way to protect the future of hunting in America.

One group (the no traps no dogs bunch) either doesn't bother to protect trapping or dog running laws because those laws don't apply to them, or they support the loss of those practices (even if only in theory), because they see them as public-relations dangers to the hunting that they do practice, and, perhaps, like the proverb goes, it's better to be the flexible green reed that bends in the wind than the mighty oak that will break in a storm. This group (who my lawyer friend Alex calls the Negotiators) hopes being flexible about some hunting regulations now will give them credibility and a reputation for being reasonable in future legislative arguments.

The second group (the trappers and dog runners, who Alex calls the Hardliners) believes that when trapping and dog running laws are chipped away, they simply represent the easiest laws with which to start–and that the chipping will never stop, and, through the loss of some practices here and there over time, a shifting baseline of acceptability will make it easier to do away with the next thing and the next thing, until all forms of hunting in America become a thing of the past. The Hardliners try to convince the Negotiators that even if they aren't trappers or dog runners, it's in their best interest to protect those practices. In the hunting community this is referred to as *guarding the gate*. Any law professor will tell you a "slippery slope" argument represents weak conjecture more than strong, evidence-based debate, but I do see where the Hardliners are coming from.

In 2021, California introduced a bill to ban all black bear hunting in the state. The sale of 31,450 black bear hunting licenses (at $51.02 a pop) in 2021 generated $1,604,579 for California's Department of Fish and Wildlife, and there was no plan to replace the conservation funding that would have been lost if bear hunting was banned.* The bill was withdrawn in less than a week, after an outpouring of opposition. More than 21,000 people signed a petition to get the bill thrown out, and it worked.

Curiously, there was no such outpouring of support for hunting bears with hounds when it was banned in California in 2012. Same goes for the state's outlawing of trapping in 2019. It's possible that hunters finally put their foot down after seeing the end of mountain lion hunting, using hounds, and trapping in their state. It's also possible that they just didn't care about trapping and those other fringe hunting practices and chose to only speak up about the issues that were most important to them. Either way, it seems like an example of how the community can and will successfully defend the types of hunting most precious to them, even in a state with some of the most restrictive hunting laws.

Hunters continue to debate whether it's a better strategy to band together in defense of practices some don't approve of or participate in, in an effort to keep detractors further from the practices they do participate in, or if they'd have more success adopting more stringent self-policing with the understanding that if they don't police themselves, someone else will. Nothing less than the future of hunting is at stake, and while that affects me as I spend more time hunting, my bigger concern is for the lack of support for evidence-based conservation from others who value wildlife. Hunters, regardless of their intentions, have stumbled upon a functional method of support for animals and their habitats. We (anyone interested

---

*This aspect of lost funding is almost never considered by anti-hunting groups who are itching to get rid of hunting before finding a conservation funding alternative that generates anything close to the same amount of money for habitats or wild animals. It would be more beneficial to the animals if the anti-hunting groups focused on raising their own two million dollars annually for the bears before they started picking on hunting. But perhaps then they'd see how our four million dollars together would be even better for the bears, and we could start working as a team on behalf of bears and their shrinking habitats. A guy can dream.

in animals enough to read this book) forget that most people on earth don't have the time, money, or inclination to bother with the finer points of conservation or philosophize how to keep animals and habitats around for the generations that follow us. It's hard enough for most folks to manage their own life, let alone the existence of animals and their habitats. We (the people who have the luxury to care about this stuff) are so close to all being on the same side, which is perhaps what makes hunter versus anti-hunter debates so frustrating.

APPENDIX I:

THE NEXT BEST THING

You may be opening a tab this very moment to check the status of Apprentice Hunting Licenses in your state. Or perhaps you're one of the 87-ish percent of Americans who approves of hunting, but you don't think it's something you have the time or inclination to pick up just yet. I like to imagine that having read this far, what you want most of all is a source of meat that's easy on animals and the environment. If you're someone who doesn't hunt or if you're a hunter who's struck out for the season and facing an empty freezer, you have some options for the next best thing.

Hunting is at the top of the environmentally friendly meat ladder—the money spent to hunt goes back to protect the environment and the animal has gotten to live its most natural life—but hunting isn't a practical way for everyone to source meat. Second to hunting, local farms and ranches that practice rotational grazing regenerative ranching are your best bet. These cattle (for example) are feeding only on grass, and they are moved from pasture to pasture so the grasses can recover and the soil can be nourished by the natural fertilizer the cattle "deposit" on the landscape. This system mimics nature's cycle, which keeps soil healthy. And healthy soil can better hold onto water and sequester and recycle carbon.

Local is the ideal way to keep transportation distances to a minimum and engage with your region's seasonal food production cycle. Talk to local providers at your farmers market or butcher shop. As younger generations take up ranching, they bring their tech skills with them, so finding and ordering high quality beef from a truly local producer has never been easier. Ask your provider if they offer shares of beef or look online for a local provider who does. If you don't mind buying in bulk, you can reserve anywhere from a one-eighth portion of beef up to a whole steer (and be set for the year if you have the freezer space). Buying this way helps farmers plan their seasons and offers the easiest way to get high-quality grass-fed beef for the lowest prices. I live in a little apartment in Brooklyn, but I have a small five-cubic-foot chest freezer that keeps me in wild game and grass-fed beef and bison all year long. Not attainable for everyone, I recognize, but far more reasonable and affordable than I thought it would be.

Further down the meat ladder we have "happy stickers" like *organic, animal welfare approved, humanely raised, pasture raised*, and so on.

Caption TK

Stickers are generally better than no stickers, but be aware there's a lot of greenwashing that goes on in the sticker-verse, and packaging is a marketer's game. If you aren't planning a full investigation of the meaning behind each symbol, take them with a grain of salt. Finding out where meat came from before it got its packaging isn't always easy, but occasionally there are clues. In his book *Springer Mountain: Meditations on Killing and Eating*, former restaurant critic and food writer Wyatt Williams describes his quest to track down the origin of a lauded "local, family farm" just north of Atlanta, only to discover what I think many well-meaning meat eaters fear. "The truck had unloaded white chickens that were all from the same place. Now they were being sorted into different brands. Some packages had the style of bare-bones budget chicken, anonymous bulk. Others described all-natural local chicken, packed in environmentally friendly earth tones. . . . They were all the same chickens. The labels just told customers what they wanted to hear, what they wanted to believe about themselves."

Which brings us to the bottom of the ladder, down in the muck, where you will find the aforementioned no-label anonymous bulk meat (if you have no clue where it came from, it didn't come from anywhere good), as well as mass-produced plant-based meat substitutes. This "cheap meat" from concentrated animal feeding operations might be easier on the wallet come check out, but it costs the most everywhere else. These costs play out in a variety of ways. The most unfortunate aspect is that "cheap" meat is the greatest contributor (meat-wise) to the climate emergency, and extreme changes in climate affect us all, right down to our wallets. Two-dollar burgers contribute to all of us having to run our air conditioners for longer, spend more on allergy medications and respiratory conditions, build flood-protection infrastructure, rebuild after wildfires, and battle worsening and novel disease from an array of vectors. Such meat should be taxed to reflect its true cost, while climate-friendly farming and ranching practices should be subsidized so more farmers will be encouraged to follow suit and more consumers could afford to buy from those producers. The fact that low-income people have no choice but to contribute to the greatest climate-change-offending-meat-production practices, which simultaneously adversely affect them the most, is an egregious failing of our food

The Cost of Meat

	Cost out of $$$$				
Meat from		Factory Farm	Small Local Farm	Regenerative Farm	Hunting
	to the Environment	$$$$	$-$$	0-$	-0 (hunting actively benefits the environment by creating value for habitat)
	to the Animals (how they live)	$$$$	$-$$	0	0
	at the cash register	$-$$	$$$	$$$$	$-$$$$+
	of Your Time	$	$$-$$$	$$-$$$	$$$-$$$$+

production regulatory systems. Meat substitutes suffer from the same system, requiring even more land and water to produce the same number of calories. Farming such large swaths of crops without animals depletes the ground of nutrients, leading to soil death. Dry, dead dirt can't hold onto water or sequester carbon. These industrialized production systems take no part in the biogenic carbon cycle making them the least environmentally friendly or sustainable.

Your money or your time, hunting will cost you a lot of one of them, at least. Therefore, it's not for everyone. Buying regeneratively raised meat doesn't cost any time, but it does cost more money than other meat you can buy. But if you have the time or the money to make one of these environmentally friendly choices, you're helping others, not just yourself.

Bison

My favorite champion of the biogenic cycle is our own national mammal, the bison. America used to be home to between thirty to sixty million bison (sixty million is a controversial and rather specious number that was estimated back in the 1800s–thirty million is closer to the more accepted, modern estimate of historical populations), and the way they move through their environment was a crucial part of Nature's annual cycle across much of what is now the United States. The big wooly tanks eat tall grasses, poop, stomp down vegetation as they graze, smoosh their poop into the ground as they walk, which aerates the soil, and roll and wallow in soil and mud. This all happens while they are constantly on the move to the next green pasture ahead of them, and it has positive impacts on the land. Their grazing on tall grasses gives a chance for other, smaller plants that don't grow as quickly to get needed sun exposure. As professor and grassland ecologist Mathew D. Moran explains, "Bison's selective grazing behavior produces higher biodiversity because it helps plants that normally are dominated by grasses to coexist." And when they eat some vegetation from over here then poop it out over there, that's spreading nutrients around the landscape and fertilizing the ground, which creates rich, robust topsoil full of living organisms and able to absorb and hold onto moisture. The plants that grow from that new vibrant soil offer habitat and nutrients to animals that share the land with bison. All those new plants and rich soil aid in carbon sequestration as a vital part of the biogenic carbon cycle. Even the divots left in the ground from bison wallowing create temporary micro-ecosystems when they fill with water. These then benefit new plant growth and insect populations, which makes the animals who eat those plants and insects thrive, as well as making the animals who eat those animals thrive. Cattle can graze the same way bison do, and when they are grazed in this manner it's called regenerative grazing or ranching.

You probably already know about the Biogenic Carbon Cycle (or as I like to call it Eat, Shit, and Die) because Mufasa gave a brief earth science lesson on the topic in *The Lion King*. "Everything we see exists together in a delicate balance. And we need to understand that balance and respect all the creatures, from the crawling ant to the leaping antelope. . . . When we die, our bodies become the grass and the antelope eat the grass. And so, we are all connected in the great circle of life." Here's a less poetic recap: Photosynthesis is the process by which plants use the sun to absorb and trap carbon dioxide. Carbon is stored in the plants, which are eaten by ruminant animals like bison and cattle. These grazing animals aerate and fertilize the soil by pooping and stomping around. Their burps and waste issue carbon as methane. Hydroxyl oxidation is the process in which that methane is transformed into carbon dioxide–it takes about ten years. Then the sun and plants grab it again, and the cycle starts over. Unlike biogenic carbon that's constantly being recycled, fossil carbon is the troublemaker that gets taken from deep within the Earth. It's freed at large quantities into our atmosphere via a one-way street.

The majority of bison, around 96 percent, are privately owned. Ted Turner manages over 45,000 bison across his ranches and opened Ted's Montana Grill, a small chain of restaurants that helped popularize bison meat as a more environmentally friendly alternative to conventional beef production. Now you can find bison meat in burger joints and grocery stores, and in jerky and protein bars, all over the country, and that's a great thing for bison. It's a "eat 'em to keep 'em" model, the same model that saw rising popularity in the early 2000s with boutique restaurants and butcher shops touting "heritage breed" pigs and chickens (as farmers chose to only raise the largest breeds of livestock, others declined in popularity–some enterprising souls kept these increasingly rare heritage breeds from going extinct by making them popular specialty meats).

More demand for bison meat means more demand for bison on the landscape. I'd love to see more ranchers trade their cattle for bison and work toward getting the American grassland prairie ecosystem back to its former glory. The ambitious conservation organization American Prairie has a simple objective, buy up land in Montana that can be connected to public land, in order to create more expansive wildlife migration corridors and protect temperate grasslands, the least protected biome on earth. The Rocky Mountain Elk Foundation often works similarly, buying land they can connect to elk migration routes. American Prairie has been so successful reintroducing bison, they've been able to help other states expand their own populations by sharing surplus animals. We may not be able to get back to thirty million bison, but we can do better than the twenty thousand we consider wild today.

Populations of grassland grouse species like prairie-chickens and sage-grouse have been declining in the West. Their habitats are overrun with invasive tall grasses and an accompanying lack of biodiversity, which has speeded their decline. Reintroduction of bison to these prairie lands would help repopulate the birds by way of clearing the tall grass and aiding the mini biomes smaller animals like the grouse need for food and breeding range. And because bison love munching the grass that grows out of recently burned ground, they help to enrich soil and plant life after fires and prescribed burns, which further helps birds that live in areas increasingly affected by forest fires. *Regenerative grazing* and *regenerative agriculture* are just the trendy buzzword names for this ancient nature cycle. The terms remind me of the area in the American Museum of Natural History that educates viewers about soil and farming practices. It's a beautiful hall with delightful wood paneling on the walls and handsome fonts above glass dioramas that proudly guide you to topics like "Decomposition" and "Fertilizers in the Soil." In a pictographic list of fertilizers, next to manure, is a box that says "Seaweed." Espousing the near magical powers of seaweed today is regarded as some next-level eco-genius farming practice—but it's not. We've been doing it in America long enough that the museum included it in their exhibit from the 1950s. We know a lot about sustainable farming practices and how to live in harmony with the land, because that's what we had to do before industrialized agriculture. Now we just need to make those practices more accessible to farmers and ranchers by offering the support they might need to transition back to these more ecologically sound farming structures.

APPENDIX 2:

HOW TO GET STARTED HUNTING

If you'd like to keep following your curiosity about hunting, I have a few suggestions about where to start. There's no perfect order. The best choice of what to do first or second will depend on where you live, the resources available to you, and how much money or time you'd like and are able to invest.

First, do you know anyone who hunts? Even through a friend or family member? Ask to be connected. See if they have any interest in taking you out sometime. Going out into the field with someone, even if you're just hanging out and observing, is the most fun and informative way to build the confidence and comfort in the woods you'll need for hunting.

Check to see if your state offers programming for new hunters on its Department of Natural Resources or Fish and Wildlife (or Fish and Game, depending on what your state calls the department) website. You should view these government-run programs as the smallest of steppingstones. Better than a how-to video but with less time and funding than they would need to be truly comprehensive.

Look up the National Deer Association and their Field to Fork program. Watch the nine-minute video they produced and see if they're offering any Field to Fork events in your area. This program (which brilliantly seeks out its new recruits at farmers markets) is more comprehensive, with more resources available to it, than most government offerings can afford, but it's not nationwide. NDA's Field to Fork program is just one specific example; there are an increasing number of programs, schools, and camps popping up to teach hunting skills to adults eager to learn. I have no doubt an internet search that includes "learn to hunt" plus your city or state will turn up a few offerings–more are being added every year.

Take your state's Hunter Education or Safety course. You might be able to complete the entire course online. Personally, I think everyone should do this first, but I can imagine it feeling a bit overwhelming if you're still on the fence about this whole hunting thing. Online options might ease your worry, however. The state of Maine found that when they moved Hunter's Ed online, they had a dramatic increase in women's participation–more women completed the program during the Covid-19 pandemic than they had in the previous five years combined. Many said the prospect of doing the course online was less intimidating than going to a class where they might be the only adult or only woman. There are also programs that teach hunting by the same groups who design the safety courses. Today's Hunter, for example, has courses about deer, turkey, and duck hunting all online for $30 each. And you can still go the apprentice license route. An apprentice license allows you to hunt with a licensed hunter or guide, without having to commit to the full course yet. It's not a "get out of Hunter's Ed free" card, though. You will have to take the class in order to start buying hunting licenses. And that's a good thing, even if it's all online.

A friend asked me if they could just hire any guide to take them out hunting for the first time, and I don't think it's the best idea, but it may not be the worst. If you are really coming up short everywhere else, here's how I'd go about it: Get your license or an apprentice license, then reach out to a local guide about a mentoring hunt. *Be clear about your experience and comfort level* as well as the equipment you own or don't own. It's not generally a guide's job to teach you how to hunt from scratch, so make sure they know that's what you're looking for and are able to accommodate you. Don't feel like you need to start by hunting wooly mammoths either; small game is a great way to get started and can represent minimal investment to the space- (apartment dwellers) or spend-conscious hunter.

Take a sporting clay lesson! I know I'm a biased sporting-clay freak, but I keep recommending it because it's crazy fun, a great way to learn about gun safety, and a chance to become comfortable shooting if you're new to firearms. Sporting clay ranges are outside, pretty, and less intimidating than rifle ranges. Look up "sporting clay range" and see if there are any near you. I recommend sporting clays over skeet or trap shooting because those two (both Olympic sports) can be a little trickier for the beginner and have more of a competitive feel, which I recommend avoiding when you're just starting out. I also cherish the little walk to the next station at sporting clay ranges. Similar to walking to the next tee in golf, it affords you a moment to chit chat with your companions or instructor while you stroll outside. But just like a golf course, sporting clay ranges take up a lot of space, so there's a chance you'll have an easier time finding a club that just offers skeet or trap.

Join a group. I recommend starting with Backcountry Hunters and Anglers. There are lots of hunting organizations out there that do great work for animals and the environment, but I've found BHA to be the most social. They're a great resource for new hunters and do awesome work to protect public lands for everyone to enjoy. I'd guess you could find a mentor with relative ease through your local chapter. You can (and should) collect memberships from other groups too, as you find your community and the specific lands and animals you feel most connected to.

Be cognizant of the time of year. A generalized view of hunting seasons looks like this:

Spring = turkey and black bear
Summer = fishing
Fall = upland birds, ducks, geese, big game: deer, elk, and black bear
Winter = upland birds, ducks, geese, small game: rabbits, squirrels

If it's summer, and you decide you want to give deer hunting a try, then you've got some time to take a lesson, meet some people, take Hunter Safety, and get in gear for a fall hunt. If it's already mid-November, then you're going to be hard pressed for time. Don't rush. If you miss deer season that's no big deal, it'll be back. Take the opportunity the extra time affords you to learn more, meet more people, and look to small game and turkey hunting first. Turkeys can make for the ideal first hunt because calling them in is exciting and they aren't furry or cute. Noble yes, but not cute.

Can you just go it alone? You can, but I don't recommend it (I say at the advice of council). If you wanted to try, here's what I'd say; Take Hunter's Ed, buy a license, read some books, watch some how-to videos, *read your state's hunting manual for that year* (they change annually), then call your local Conservation Officer if you have any questions. In 2022, the Senate passed the bipartisan MAPLand Act (Modernizing Access to our Public Land) so government mapping of public lands could become digitized and therefore more accessible to people using GPS-enabled devices like smartphones. This makes it much easier to find public hunting land near you. If you want more features than just public land boundaries, onX makes an easy-to-use mapping program I've found helpful and had a lot of fun using. When you head into the woods, start with small game to hone your skills. I don't recommend starting out alone if you have other options, but it's not impossible.

Time and Money

Time and money are usually the greatest barrier to entry for new hunters. Hunting costs you both, but don't let that spook you. You don't need a bunch of fancy gizmos or expensive outdoor gear from elite hunting-specific brands. Plenty of people have brought home wild critters for dinner after a day of hunting in jeans (though, as we used to say on the river, "cotton is rotten," so if you're in an environment where you might get wet, stick to items that dry quickly or keep you warm even when sodden). The clothing you need will be dictated by where you are and the type of hunting you're doing, just like any outdoor wear. If you have a mentor and you're going out together, maybe they can lend you a few things for your first days in the field.

After a few days of hunting, I bet you'll get the bug and want to invest in some of your own stuff. Scour thrift stores, Army Navy surpluses, and eBay first. I scored big at a Goodwill in Montana once. You don't have to jump right into the deep end if you don't want to. Spending time outside is more important than what you're wearing while you do it. It's okay to wait and learn from others while you make your own mistakes–then you'll know the stuff you need to spend money on the most. Try to remember that people have been hunting throughout human history without whatever the hot gear of today is. As outdoor philosopher Yvon Chouinard says, "The more you know, the less you need."

Depending on where you live, you could be squirrel hunting your local public land in the time it takes to get your Hunter's Ed certification and for the cost of a .22 rifle, say $300, and a box of rounds, say $15 (and don't forget that a big chunk of that goes back to conservation from excise taxes and Hunter's Ed fees). The more serious you get about hunting, the more you'll probably end up spending. Same as if you got really into model railroads.

As far as time goes, there's no way to cut corners. The more time you spend in the woods, the more chances you'll have to see animals and learn new things. I've never been excited for an early flight out of JFK, even if I knew that flight was taking me someplace I was excited to go. I have however, been excited to wake up even earlier to go hunting, despite having no idea if I'd be "successful" or not. Between the planning and the scouting and the time you spend actually hunting, it can take up a lot of your time. But the more time you spend scouting and practicing, the more likely you are to be successful, and I bet you'll only be upset that you can't spend more time doing it. Renewed or newfound interest in hunting and spending time outside has spurred many Parks and Wildlife Services as well as private organizations to implement community programing that offers more mentorship match-making programs. Support from USFW, local clubs, and national organizations can help you navigate your financial and time concerns.

One of the many things that separates hunting in America from other countries is just how egalitarian it is. When someone hears "hunter" in America they might conjure up the image of a rich person just as quickly as they might think of a poor person. The United Kingdom still suffers from classist views on hunting from the days it was only accessible to nobility. But in the States hunting is enjoyed by every socioeconomic stratum and across the political divide. No matter who you are or what you spend, you'll never eat better for less.

Kids

Despite the time required for hunting and parenting, I have friends who manage to be both hunters and parents quite competently. And even though I'll never be someone's parent, I'm definitely someone's kid, and I can tell you how much I cherish the memories I have of the time I spent outside with my dad fishing, shooting, and horseback riding. You can continue to hunt and fish into old age, so these pursuits make for a great way to spend time together no matter how old you (or they) are. I think it's ideal to have a shared interest in an activity with your kids. And if that activity creates a respect for nature and puts food on the table you can share together, all the better. I'd recommend reading *Stolen Focus* by Johann Hari (regardless of your kid-status), *Last Child in the Woods* by Richard Louv, and *Outdoor Kids in an Inside World* by Steven Rinella if you find the intersections of childhood and nature interesting.

Fun

Yes, hunting and fishing is fun. Sometimes it's hard to admit that. *Exciting* or *thrilling* are perhaps the synonyms more commonly used, and descriptions change based on the pursuit and when it took place. For instance, hunting birds like ducks and pheasant is fun while you're doing it. You can do it with friends, and you can talk a bit. I would never describe deer hunting as fun while I'm doing it. I do not like getting up well before dawn to freeze my ass off all day. But it is exciting. Only after I arrive back home—even if I've returned empty handed—would I call it fun. It must be, or I wouldn't wake up early the next day to do it all over again (or maybe it's the operant conditioning of a variable-ratio success schedule).

It's good that hunting and fishing are fun. Fun means people keep doing it. The more people who have fun outdoors, the more people there are who want to keep the outdoors fun. That means clean air for hiking, clean water for healthy aquatic life, lots of critters to photograph and hunt, and healthy habitats for them to live in. The government can sell hunting and fishing licenses, but if nobody is buying them, then they can just as easily sell drilling and mining rights. So go have fun outdoors and pay to do it. Nature is not free.

RECOMMENDED READING

Some of the following works served as sources. Others I just wanted to share. Not all my sources are listed here, but these are the titles I'd recommend if you want to continue reading about some of the topics I covered.

For the Bison Nerd

Theodore Roosevelt & Bison Restoration on the Great Plains by Keith Aune and Glenn Plumb
(Short and sweet if you want an overview.)

Mr. Hornaday's War: How a Peculiar Victorian Zookeeper Waged a Lonely Crusade for Wildlife that Changed the World by Stefan Bechtel

The Extermination of the American Bison by William T Hornaday
(This is deep nerd level if you want to hear it right from Hornaday.)

American Buffalo: In Search of a Lost Icon by Steven Rinella
(If you just want one bison book, or just enjoy natural history, this one covers it all.)

For the History Buff

The Great Naturalists edited by Robert Huxley
(From Aristotle to Darwin there's a brief bio of 39 of the earliest naturalists.)

Fur, Fortune, and Empire: The Epic History of the Fur Trade in America by Eric Jay Dolan
(Any and all Eric Jay Dolan will provide a rich cultural and natural history.)

Eager: The Surprising, Secret Life of Beavers and Why they Matter by Ben Goldfarb
(You can find a species-specific natural history on any animal you want. I just love beavers. Like bison, passenger pigeons, and wolves, they have an interesting cultural story.)

A Feathered River Across the Sky: The Passenger Pigeon's Flight to Extinction by Joel Greenberg
Heart and Blood: Living with Deer in America by Richard Nelson
Yellowstone and the Smithsonian: Centers of Wildlife Conservation by Diane Smith
(Short and to the point, good for "museum people.")
Mr. Peale's Museum: Charles Willson Peale and the First Popular Museum of Natural Science and Art by Charles Coleman Sellers
(If you're a museum fanatic.)
Beloved Beasts: Fighting for Life in an Age of Extinction by Michelle Nijhuis
(An elegant bio of multiple major conservation players who defined the modern movement.)
The Father of American Conservation: George Bird Grinnell Adventurer, Activist, and Author by Thom Hatch
Natural Rivals: John Muir, Gifford Pinchot and the Creation of America's Public Lands by John Clayton
Gifford Pinchot and the Making of Modern Environmentalism by Char Miller
Aldo Leopold: His Life and Work by Curt Meine
(If you really like biographies.)

For the Taxidermy Enthusiast

Nature's Mirror: How Taxidermists Shaped America's Natural History Museums and Saved Endangered Species by Mary Anne Andrei
(For those who think taxidermy is only a head on a wall.)
Still Life: Adventures in Taxidermy by Melissa Milgrom
(A fun, broad look at taxidermy.)
The Authentic Animal: Inside the Odd and Obsessive World of Taxidermy by Dave Madden
Kingdom Under Glass: A Tale of Obsession, Adventure, and One Man's Quest to Preserve the World's Great Animals by Jay Kirk
(The Carl Akeley book.)

Charles Waterton (1782-1865) and his Eccentric Taxidermy by P.A. Morris
(A hoot. All of Dr. Pat Morris' work is fun.)
Walter Potter's Curious World of Taxidermy by Dr. Pat Morris with Joanna Ebenstein
Best Friends Forever by J.D. Powe
(A delightful photography book of preserved dogs and their history.)
Reflecting the Sublime: The Rebirth of an American Icon by Douglas Coffman
(This short book is just the history of one bison diorama made by Hornaday.)
Stuffed Animals and Pickled Heads: The Culture and Evolution of Natural History Museums by Stephan T. Asma
Painting Actuality: Diorama Art of James Perry Wilson by Michael Anderson
(A grand biography that joins diorama art, taxidermy, natural history, and the importance of museums.)

For the Teddy Addict

The Wilderness Warrior: Theodore Roosevelt and the Crusade for America by Douglas Brinkley
(If you're really into Teddy, conservation, and reading giant books.)
The Naturalist: Theodore Roosevelt, A Lifetime of Exploration, and The Triumph of American Natural History by Darrin Lunde
(If you want the short and sweet portable version.)
The Rise of Theodore Roosevelt by Edmund Morris
Camping and Tramping with Roosevelt by John Burroughs
(A joy to keep on you while spending time outside.)

Conservation in Africa

"Trophy Hunting Bans Imperil Biodiversity" for *Science 365* (6456) by Amy Dickman, Rosie Cooney, Paul J Johnson, Maxi Louis Pia and Dilys Roe (plus 128 Signatories)
(If you're interested in functional conservation in Africa, track down

as much as you can find from Amy Dickman and Rosie Cooney. Scientists who are non-hunters but pro-hunting, because they see it work.)

Game Changer: Animal Rights and the Fate of Africa's Wildlife by Glen Martin

The Myth of Wild Africa: Conservation Without Illusion by Jonathan S. Adams and Thomas O. McShane

"Why Zambia Lifted the Ban on Hunting Lions and Leopards" for BBC Africa by Milton Nkosi

Poached: Inside the Dark World of Wildlife Trafficking by Rachel Love Nuwer

"The Rhino Hunter" for *Radiolab* by WNYC Studios

"The Economic Impact of Trophy Hunting in the South African Wildlife Industry" from *Global Ecology and Conservation* by Melville Saayman, Petrus Van der Merwe, and Andrea Saayman

"Conservationists Should Support Trophy Hunting: Why I Joined 132 Researchers in Signing an Open Letter in Science Magazine" for the Property and Environment Research Center by Catherine E. Semcer

Operation Lock and the War on Rhino Poaching by John Hanks

Modern Huntsman, Volume Eight: Africa
(A handsome edition that covers a range of topics from a variety of sources. Good for the newcomer who doesn't want to read a whole depressing book about poaching and habitat loss.)

Food and the Environment

To Boldly Grow: Finding Joy, Adventure, and Dinner in Your Own Backyard by Tamar Haspel
(Haspel gets into hunting, but this book covers the gambit of DIY food; what's reasonable, fun, and worth it, while acknowledging what is and isn't beyond the scope of most people to achieve. She did the homework for you.)

Defending Beef: The Ecological and Nutritional Case for Meat by Nicolette Hahn Niman

Sacred Cow: The Case for (Better) Meat: Why Well-Raised Meat is Good for You and Good for the Planet by Diana Rodgers RD and Robb Wolf

Springer Mountain: Meditations on Killing and Eating by Wyatt Williams
The Unsettling of America: Culture and Agriculture by Wendell Berry
Eating Aliens: One Man's Adventures Hunting Invasive Animal Species by Jackson Landers
The Omnivore's Dilemma: A Natural History of Four Meals by Michael Pollan
The Scavenger's Guide to Haute Cuisine: How I Spent a Year in the American Wild to Re-create a Feast from the Classic Recipes of French Master Chef Auguste Escoffier by Steven Rinella
Hunt, Gather, Cook: Finding the Forgotten Feast by Hank Shaw (This is a cookbook, but with each introduction you see why chef Hank Shaw wanted to focus on wild foods for himself and the environment.)

Conservation Funding and Management

A Sand County Almanac: With Essays on Conservation from Round River by Aldo Leopold
"Hunting is Slowly Dying Off and That Has Created a Crisis for the Nation's Many Endangered Species" for *The Washington Post* by Kyle Grantham
Modern Huntsman, Volume Three: Wildlife Management (Another good generalist's look at hunting, fishing and their connection to conservation and healthy ecosystems.)
The Duck Stamp Story: Art, Conservation, History by Eric Jay Dolin and Bob Dumaine
The Wild Duck Chase: Inside the Strange and Wonderful World of the Federal Duck Stamp Contest by Martin J. Smith
Nature Wars: The Incredible Story of How Wildlife Comebacks Turned Backyards into Battlegrounds by Jim Sterba
The New Economy of Nature: The Quest to Make Conservation Profitable by Gretchen C. Daily and Katherine Ellison
Nature's Fortune: How Business and Society Thrive by Investing in Nature by Jonathan S. Adams and Mark R. Tercek

Hunting Philosophy and Ethos

Call of the Mild: Learning to Hunt my Own Dinner by Lily Raff McCaulou

Meat Eater: Adventures from the Life of an American Hunter by Steven Rinella

(*Call* and *Mindful Carnivore* are both great why-I-started-hunting-as-an-adult stories, but if you've had enough of that perspective for now, read Meat Eater next and hear the perspective of the lifelong hunter.)

The Mindful Carnivore: A Vegetarian's Hunt for Sustenance by Tovar Cerulli

Beyond Fair Chase: The Ethic and Tradition of Hunting by Jim Posewitz

(Jim Posewitz is a hero in the hunting and conservation community. He passed away in 2020 at 85.)

A Hunter's Heart essays collected by David Petersen

Heartsblood: Hunting, Spirituality, and Wildness in America by David Petersen

The Hunter: Developmental Stages and Ethics by Dr. Bob Norton

In Defense of Hunting by James A. Swan

New Adult Hunters

"Why Women Are the Fastest Growing Segment of the Population Who Hunt" for *Forbes* by Chris Dorsey

"Is a New Breed of Hunter Joining Our Ranks?" for *Grand View Outdoors* by Amy Hatfield

"Hipsters Who Hunt: More Liberals are Shooting Their Own Supper" for *Slate* by Emma Marris

"Put Down the Kombucha and Pick Up a Crossbow: Hipsters Are the New Hunters" for *The Wall Street Journal* by Cameron McWhirter and Zusha Elinson

"All the Cool Girls Hunt their Own Food" for *Jezebel* by Erin Gloria Ryan

"More Women Give Hunting a Shot" for *National Geographic* by Kristen A. Schmitt

"A New Breed of Hunters Focuses on the Cooking" for *The New York Times* by Kim Severson
"So You Want to Be a Hunter" for *Outside* by Wes Siler
Modern Huntsman, Volume Four: The Women's Issue

Outdoor Cats

"That Cuddly Kitty is Deadlier Than You Think" for *The New York Times* by Natalie Angier
"The Moral Cost of Cats" for *Smithsonian Magazine* by Rachel E Gross
"'Caught by Cats' Photo Sheds Light on Cats' Killing of Birds" for *National Geographic* by Cordilia James
Cat Wars: The Devastating Consequences of a Cuddly Killer by Peter P. Marra and Chris Santella
"The Killer at Home: House Cats Have More Impact on Local Wildlife than Wild Predators" for *NPR Science* by Lauren Sommer

People and Other Animals

Wild Souls: Freedom and Flourishing in the Non-Human World by Emma Marris
(A wonderful compendium of how humans relate to other animals by the renowned nature writer.)
Fuzz: When Nature Breaks the Law by Mary Roach
(A delight, like all of Mary Roach's works. I've heard criticism that she doesn't get into the deeper environmental issues that stem from the stories she chooses to tell, but that's not the point of the book. If you're curious about synanthropes and human-wildlife conflict, then there's lots of books on those topics to pick up afterward.)
Wild Ones: A Sometimes Dismaying, Weirdly Reassuring Story About Looking at People Looking at Animals in America by Jon Mooallem
Where the Deer and the Antelope Play: The Pastoral Observations of one Ignorant American Who Loves to Walk Outside by Nick Offerman
The Medici Giraffe: And Other Tales of Exotic Animals and Power by Marina Belozerskaya

ACKNOWLEDGMENTS

I owe so much to the authors and environmental advocates whose heavy lifting make my work easier. Books are the greatest form of sanctioned cheating. In just 200 or so pages, I can gain all the knowledge from a project that took someone years or a lifetime to complete. I'm even more thankful to those who took time away from their own work to help me with mine. That said, I wouldn't be able to thank anyone if it weren't for Will McKay at Timber Press, who has been so kind and supportive of me during my author growing pains. And my agent at Squid Ink, Amy Collins for all the hand holding, and illustrator Sara Becan who introduced me to her. But my editor Cobi Lawson deserves the most credit and thanks. She loves cats and hiking and still went to bat for me every day while she turned my writing into a book.

A few folks actually did some writing for me—you may not see them mentioned in the text, but they were valuable resources none the less. Thanks to Kris De La Torre for always saying yes; Nicole Pursell for a missing essay on eating habits; Adam Mullins-Khatib for always being awake when I want to talk about movies; Kevin Green for the lesson in government-assisted farming programs; and Fisher Neil for being a good sport and hunting pal.

To those I couldn't coerce into writing for me, thank you for spending hours on the phone or Zoom, while I scribbled notes and asked you to repeat yourself ad nauseum. Thanks to lawyers and friends Alex Blaszczuk and Ryan Micallef who made for great sparing partners; economist Peter Reuter for the thought experiments; Ret Talbot for the three-hour fish extravaganza I regretfully had to whittle down to a soundbite; Emma Marris for her inspiration and accessibility; Sophie and Riley Egan for their stories; Tim Shinabarger for his friendly chats, incredible eye, and glorious moustache; author SJ Keller and taxidermist Travis de Villiers for discussing sensitive subjects; and the *We Hate Movies* gang, Stephen Sajdak, Eric Szyszka, Chris Cabin, and Andrew Jupin for telling me which members of the *Predator* team they would most like to eat.

I was lucky to talk with representatives from various organizations who were equally generous with their time and expertise. Thanks to: Sara Wilson and Rowlands Kaotcha for their stories from the Hunger Project; Mark Staples from the British Game Assurance; Adam Gall from Timber to Table Guide Service; Robbie Kröger and the Blood Origins team; Kevin Hurley from the Wild Sheep Foundation and Bruce Rich from the Rocky Mountain Elk Foundation for sitting down with me at their rad HQs; Matt Lindler for being so enthusiastic about dino birds at the National Wild Turkey Federation; Hank Forester for still taking my calls at the National Deer Association; the handful of people who walked me through some legal history at the Congressional Sportsman's Foundation; and all the organizations that fight for land and water so it can be given back to wildlife.

I also owe thanks to the people who helped me write this thing with and without their knowledge: Matt McNamara for weighing my meat; Kyle Munkittrick for helping to plot the trickiest of charts; Dr. Natalie Stemati who took time away from her practice to teach me how to use Microsoft Excel; Lily Lamboy and Cloe Shasha Brooks for feedback sessions. And to the wonderful people who continue to give me writing fodder, including Colleen and Mike Barcone and all the fine folks at West Kill Brewing, who deserve a rent check but get this instead; John Davison for always being my lead council, Chris Church for taking me on my first hunting trip; Jodie Errington and Andrew Marvel, constant future hunt buddies; Ben Turley for showing me how the sausage gets made; and the Morbid Anatomy crew, Joanna Ebenstein, Laetitia Barbier, and Daisy Tainton, may you never run out of curiosity. And thanks finally to my dad, who taught me how to ride, shoot, and fish, and told me I could take those skills anywhere I wanted.

PHOTO & ILLUSTRATION CREDITS

TK

INDEX

TK

Index

Index

Index

Index

Index